The Thin Bone Vault

The Origin of Human Intelligence

The Thin Bone Vault

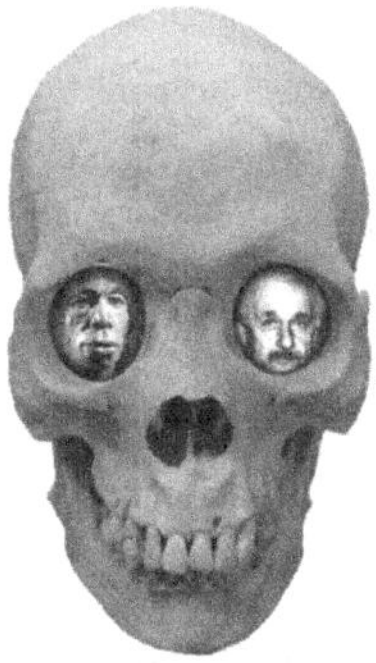

The Origin of Human Intelligence

Fredric M. Menger
Emory University, USA

Imperial College Press
ICP

Published by

Imperial College Press
57 Shelton Street
Covent Garden
London WC2H 9HE

Distributed by

World Scientific Publishing Co. Pte. Ltd.

5 Toh Tuck Link, Singapore 596224

USA office: 27 Warren Street, Suite 401-402, Hackensack, NJ 07601

UK office: 57 Shelton Street, Covent Garden, London WC2H 9HE

Library of Congress Cataloging-in-Publication Data
Menger, Fredric M., 1937-
The thin bone vault : the origin of human intelligence / Fredric M Menger.
p. cm.
ISBN-13: 978-1-84816-336-2 (hardcover : alk. paper)
ISBN-10: 1-84816-336-3 (hardcover : alk. paper)
1. Intellect. 2. Brain--Evolution. 3. Human evolution. I. Title.
QP398.M26 2009
612.8'2--dc22
2008053871

British Library Cataloguing-in-Publication Data
A catalogue record for this book is available from the British Library.

Typeset by Stallion Press
E-mail: enquiries@stallionpress.com

Printed in Singapore.

"Here is the skull of a man: a man's thoughts
and emotions have moved under the thin bone vault..."

From *De Rerum Virtue* by Robinson Jeffers

CONTENTS

ACKNOWLEDGMENT

Dr. Eva Menger assisted in the concept of the book cover. And during the years it took to write this book, my wife Lib sustained me with her affection, gardening, and Southern cooking.

Section 1

EVOLUTION

Chapter 1

INTRODUCTORY REMARKS

My original plan for this book was simple: I, a chemist with a long-standing involvement in biological problems, would examine and assess the current views on evolution in as impartial a fashion as possible. I became interested in this project when, in the course of a few hours' reading on the subject, I encountered vehemently opposed opinions by seemingly reasonable people. Many supported Darwinian ideas:

> Our own existence once presented the greatest of all mysteries, but...it is a mystery no longer because it is solved. Darwin and Wallace solved it.
>
> The Darwinian theory of evolution is the great, global, organizing principle of biology.
>
> The search for new approaches does not mean that natural selection is to be overthrown. The core of neo-Darwinist synthesis will remain valid.
>
> Except for those skeptics who are willing to discard rationality, Darwin's theory has now become Darwin's law.
>
> Darwin basically told us all we know and all we need to know about life.

Others, on the other hand, took a rather different position:

> Darwinism is a theory that has been put to test and found false.
>
> Although selective filtering and enhancement of useful genes could obviously occur in the Darwinian manner for individual cases, the explanation has the flavor of a just-so story. It is far more difficult to demonstrate that there will be a systematic accumulation of myriads of such changes to produce a coherent pattern of species advancement.
>
> Self-organizing behavior rather than selection is responsible for evolution.
>
> The transformation of masses of populations by imperceptible steps guided by selection is so inapplicable to the fact that we can only marvel both at the want of penetration displayed by the advocates of such a proposition, and at the forensic skill by which it was made to appear acceptable.
>
> The explanatory doctrines of biological evolution do not stand up to an objective, in-depth criticism. They prove to be either in conflict with reality or else incapable of solving the major problems involved.

None of the pro-Darwinian statements was written by a person with an obvious axe to grind. And none of the anti-Darwinian statements was written by a person with a religious agenda. Quite the contrary, all the quotes come from established experts in the field of biology and evolution (whose names have not been identified here in order to avoid irrelevant personality issues). The point is that there exists a scientific dilemma of major scope, and it became a personal goal to identify the source of the problem. In the process of realizing this goal, a new view of evolution with widespread implications was developed, and thus did this book come into existence.

Researching a contentious field requires a broad awareness of the subject coupled to an open mind. My acquiring an expertise in evolution was accomplished by reading and absorbing a small library of books and articles, both old and new. Many of these are listed in the bibliography. But I found that books and articles on evolution were insufficient. Since evolution is such an all-encompassing field (embodying anthropology, archeology, ecology, genetics, geology, molecular biology, paleontology, sociology, and statistics), it was necessary to delve into these subjects as well. This was both an easy and difficult task depending upon how one looks at it. Compared to the esoteric and mathematical concepts found in my main area of expertise (chemistry), evolutionary science seems relatively tractable. Many great ideas in biology, including evolution, can be explained in a simple and palatable manner. On the other hand, I encountered a bewildering array of facts and observations. I had to, therefore, search through an enormous amount of material for the information I needed and trusted (a bookish counterpart to an archeological dig!). Important ideas had to be separated from trivial ones; established truths had to be distinguished from conjecture; and valid arguments had to be differentiated from specious assertions. It was the sheer mass of information, and the need to compile and condense it, that presented the challenge.

There has, of course, also been an information explosion in my own field of chemistry (as evidenced by the excess of 100 000 chemistry articles per year from the United States alone). Chemists engaged in basic research must, therefore, confine themselves to a tiny sub-specialty and, even so, they manage to read only a fraction of their sub-specialty's output. Time is too short to do otherwise. The writing of this book was pure joy because I could escape this scientific constraint. I learned of animal life I never knew existed; of adaptations that amuse and confound; of humanoids long gone; of brilliant experiments in genetics; of the mind and brain; of language and culture. In short, as a result of my research for this book, I have developed an awareness and appreciation for Nature that a long and tortuous educational system had never imparted to me. From a selfish point of view, my new-found wonder has already richly rewarded me for my time and effort. If some of this wonder carries over to others, my rewards will be compounded even further.

Jacques Monod wrote that in the beginning, "the Universe was not pregnant with life nor the biosphere with man." From where, then, did the current diversity of life arise? According to fossil and other evidence, life began as single-celled animals and increased in complexity until advanced species, such as man, populated the earth. The process is called evolution or, in Darwin's terminology, "descent with modification". Proof of evolution is plentiful but will not be presented here; I will simply operate under the premise that evolution is an established fact. Just how evolution occurred is another matter entirely. As already pointed out, the mechanism of evolutionary change is a highly controversial issue even among serious scholars. Thus it is the mechanism of evolution, not its reality, that is the main focus of this work.

The book begins with a review of Darwin's theory of natural selection, the intellectual underpinning of all modern discussions of evolution. It was necessary to write this chapter in some detail because many of the criticisms of the theory stem, it seems, from misunderstandings. In the next chapter, entitled "Darwin Analyzed", I confront and dismiss a wide variety of criticisms that have been levied against natural selection. The purpose here is not only to correct certain misconceptions but also to use the misconceptions and their alterations as a vehicle for attaining a greater appreciation of Darwin's ideas. Several remarkable examples of natural history, portrayed in as simple and non-technical terms as possible, are included in this section of the book.

Up to this point the book might appear to have been written by a die-hard apologist for Darwinian thought, but such is not the case. Ultimately, in my overview of evolutionary science, I encounter at least one vitally important trait that is not explainable by natural selection alone: human intelligence. Something is seriously lacking when natural selection is invoked to rationalize the capabilities of the human brain, that wonderful organ housed in the "thin bone vault". Thus, a great deal of space will be devoted to human intelligence (an interesting subject in its own right) and to why natural selection cannot adequately explain why humans are so smart.

At the very end of the book I propose an alternative to natural selection. Its purpose is to demonstrate that natural selection is not the only possible route to biological change. Other mechanisms, however speculative, are worthy of consideration. This final section of the book draws upon modern genetics. But whether technically trained or not, the reader will acquire, I hope, an enhanced appreciation for the origins of the human mind and for humanity itself (including possibilities for the future development of our species).

Note that I do not state, and will never state, that natural selection is false. There is a crucial difference between claiming that a theory is false and claiming (as I do) that a theory must be expanded to encompass certain complexities of Nature. Modification rather than elimination of established concepts in science happens all the time. For example, the theory of disease based on bacteria was not discarded when viruses were discovered. Instead, the theory of disease was broadened to accommodate the new information.

Many people who do not work in the field of science, and some who do, have an almost innate dislike for certain aspects of evolution. These persons should take no comfort in the inadequacies of natural selection as delineated herein. My arguments are, I hope, merely a prelude to an even better scientific description of life. No serious book on evolution, including this one, can be written without standing on the shoulders of Charles Darwin. And if, while doing so, his beard is tweaked a little (or even a great deal), that is the way science works and progresses.

It is now necessary to confront, and permanently set aside, the conflict over evolution between science and religion. Many thoughtful people, I presume, wish that science and religion were at peace. Science is theologically neutral; it depends upon observation and experimentation; and it accepts as little as possible on faith. Religion, on the other hand, is a system of belief; it is not amenable to experimental testing; it addresses issues of morality and values on which science has nothing to say. The two domains are time-honored but completely different. They are pursued for different reasons. They serve different functions. And we need them both as the following anecdote illustrates.

The Northern Lights are among the most magnificent of all natural phenomena, filling the sky with a wondrous display of reds, yellows, greens, and blues. Flickering lights and colors take the shape of arcs, streamers, and great hanging tapestries. Scientists believe that Northern Lights are caused by high-speed electrons from storms on the sun. These electrons become trapped in the earth's so-called "Van Allen radiation belt" from where they are drawn to the polar regions by the planetary magnetic fields. Indians of northern America, however, have a different interpretation of Northern Lights; they believe that the lights are the campfires of dead ancestors. Who is right — the scientist or the Indian? It depends.

When I am camped at the edge of a northern lake on a clear winter's night, and I gaze up into the sky at the Northern Lights, I am unable to think of electrons trapped in the Van Allen belt. I need to believe that I am standing where a race of ancient people had camped before. I need to believe that they are still thriving and are sending back to earth evidence of their continued existence. Believing this makes me happy and fulfilled.

But back in the laboratory, I might have to deal with the alternate scientific viewpoint, and I could do so with enthusiasm. If all this seems schizophrenic, blame it on whatever in us that demands both non-testable notions and scientific facts lying side-by-side. Albert Einstein said it right: "A legitimate conflict between science and religion cannot exist. Science without religion is lame, religion without science is blind." Sir William Bragg's famous dictum should also be cited: "Religion and science are opposed...but only in the same sense as that in which my thumb and forefinger are opposed...and between the two, one can grasp everything." With these two wonderful quotes, I permanently lay to rest any further discussion of religious issues.

For whom, then, is this book intended? The format is geared to the non-professional readership in that I minimize jargon and unnecessary complexities. Accordingly, I join DeKruif, Gould, Huxley, Medawar, Schröedinger, and Wells (to immodestly name a few) who also wrote popular books on biology. Perhaps the style of this book has been most influenced by E. Schröedinger, of quantum mechanics fame, who stated: "If you cannot — in the long run — tell everyone what you have been doing, your doing has been worthless." But despite having set my sight on the non-expert, I prefer not to let evolutionists off the hook entirely. It might not be a bad exercise for evolutionists (that strange mix of overly strident supporters and overly critical detractors of Darwin) to peruse the book as well. After all, it has been written by an "outsider" whose independence from evolutionist cliques might not guarantee an impartial perspective, but it certainly cannot hurt. And, as mentioned, toward the end of the book I propose alterations of natural selection to account for those cases, such as the contents of humans' thin bone vault, where Darwinism seems to fail. This material is new, fresh, and provocative even to the expert.

In his preface to "*The Sense of Beauty*", philosopher George Santayana wrote the following words:

> The influences under which the book has been written are rather too general and pervasive to admit of specification; yet the student of philosophy will not fail to perceive how much I owe to writers, both living and dead, to whom no honor could be added by my acknowledgements. I have usually omitted any reference to them in footnotes or in the text in order that the air of controversy might be avoided, and the reader might be enabled to compare what is said more directly with the reality of his own experience.

Although I could echo a similar sentiment, I will depart from Santayana's approach by listing (more in the style of a text than a treatise) most of my sources in a bibliography. These sources must not be blamed for my comments, speculations, and (especially) my theorizing that lie at the heart of this book. The bibliography also contains books that are recommended for further reading.

Obviously, it was not possible to verify first-hand all the natural history and the myriad of other details presented to enliven this book. I can only assure the reader that I have done my best to search out material from seemingly reliable sources. Although some will, no doubt, be able to quarrel with points made here and there (debates in the fields of evolution and anthropology abound), I hope that these will not be employed to discredit my major themes which, fortunately, are never dependent upon the validity of any single observation or upon the precision of a particular number. For example, for the purposes of this book it makes little difference if a human trait first appeared 30 000 years ago or 50 000 years ago (although this might be considered a serious uncertainty in anthropological circles). Thus, the sweep of the book far exceeds the particulars.

Chapter 2

DARWIN AND NATURAL SELECTION

There are two principal ways in which knowledge expands: (a) through careful gathering of facts via observation or experimentation and (b) through a flash of insight that elevates the world to a new level of understanding. In 1859, Charles Darwin published a book, entitled "*The Origin of Species*", that incorporated both of these. Thus, astute observations on finches and other animals in the field led Darwin to verify the fact that, over time, organisms do indeed change ("evolve"). The idea of evolution was, incidentally, not new to Darwin. Decades before him people like Lamarck had also contemplated the "march of Nature". In fact, Darwin's own grandfather, Erasmus, wrote an immensely long poem, with the unlikely name of "*Zoonomia*", in which evolution was a central theme. The flash of insight came when Darwin explained this evolution by an amazingly simple and ingenious mechanism called "natural selection". This original concept rests on three premises:

a) Inheritable differences crop up spontaneously and randomly among members of a species.
b) In a highly competitive world, those possessing a favorable trait with respect to the demands of the environment are more likely to survive and reproduce than those without the trait.
c) The favorable trait will be passed on preferentially to the offspring, thereby perpetuating the trait and increasing its

frequency in the population. Gradually, through a series of minute changes over an almost unimaginably long time, species will evolve and new species will appear.

The arctic hare can be cited as an example. A small fraction of hares might have a slightly thicker fur than the majority of the population. If winters should become colder than normal, hares with the thicker and warmer fur will, on the average, survive in greater numbers than those less well endowed. The thick-fur trait is passed on to the descendants and, over time, the entire population attains a heavier coat. In this manner, a thicker-furred hare will evolve by natural selection.

Since Darwin lived before the advent of genetics, he did not realize that random mutations were at least one source of his variations. A modern formulation of natural selection that embodies genetics is now called "neo-Darwinism" or the "neo-Darwinian synthesis". Although neo-Darwinism is a more sophisticated construct than that originally set forth in "*The Origin*", natural selection remains at the core of current evolutionary thought. To restate Darwin's idea in modern terms: Natural selection is a process whereby, in the continual struggle for resources, badly adapted mutants and other genetic variants compete poorly and die. A mutation that represents an advantage, however, will ultimately become disseminated throughout the population, leading to evolution of the species.

Although random mutational variations of genes are usually cited as the main source of organism variation, it must be realized that there are other proposed mechanisms for genetic novelty. These include genetic drift, sexual selection, symbiogenesis (the exchange of genes between two organisms living in close physical contact), the presence of stress proteins that increase mutation rates, the exploitation of recombination hotspots, and gene amplification. Indeed, many sources probably overstate the contribution of random mutations per se to the evolutionary process. Delving into the details of multiple genetic variation mechanisms is, however, beyond the scope and needs of this book. Suffice it to say that

I will sometimes use the term "random mutational variation" merely as a generic term to encompass all routes to changes in an organism's genetic makeup.

Natural selection has often been equated with "survival of the fittest". I will avoid using this phrase because, for one thing, survival is a necessary but insufficient condition for change. For example, a long-lived but sterile animal is, evolutionarily speaking, inconsequential regardless of its longevity. It is, in fact, reproductive advantage rather than survival per se that really counts. Reproductive advantage can take rather bizarre forms as illustrated by the male dragonfly. This insect has, projecting from its penis, a whip-like prong that attaches to the mass of sperm within a previously mated female. The male dragonfly removes his rival's sperm and thereby obtains a "reproductive advantage" (at least over a variation that did not happen to have a prong).

Another problem with "survival of the fittest" relates to a criticism that the phrase represents a "tautology" (i.e. a phrase that contains no information such as "my father is a man"). Since survival occurs among the fittest, and since the fittest are those who survive, there is (ostensibly) something unpleasantly circular about "survival of the fittest". In actual fact, the criticism seems a bit unfair as can be demonstrated again with the arctic hare. The fittest hares are defined as those with the heaviest coat of fur (not as those who survive). The heaviest coats, in turn, lead to preferential survival in severe winters. This seems reasonable, logical, and informative. Nonetheless, I will not use "survival of the fittest" because it has been tainted over the years and because "natural selection" works just fine.

Cosmic rays, viruses, chemical agents, sunlight, etc. can cause mutations, i.e. chemical alterations somewhere in the DNA of the chromosomes. Mutations, and the physical changes in the organism that accompany them, are random in the sense that one cannot predict which particular gene will be modified. A cosmic ray, for example, can hit the DNA anywhere. It is not surprising, therefore, that the vast majority of mutations are harmful to the organism (just as a nail pounded randomly through a computer would, in all likelihood, result in damage). On rare occasions, however, a mutation will impart a useful trait. It is this particular type of beneficial mutation, unlikely though it might be, that at least partially drives the evolutionary process. As mentioned earlier, additional mechanisms provide for genetic variation include the incorporation of viral and bacterial genes into host cells. Accordingly, species can also evolve by inheritance of acquired genomes rather than exclusively by mutational changes.

Many people, even great scientists, have incorrectly regarded natural selection as an absurd "adaptation by chance". For example, Nobel Prize winner A. Szent-Gyorgi wrote: "Random shuttling of bricks will never build a castle or Greek temple, however long the available time." Another popular and derisive analogy to natural selection is that of a tornado tearing through a junkyard to create, through a random assembly of scraps, an airplane. Still another describes a blind watchman who hand-builds an intricate clock (the blind watchman corresponding to Nature, the clock to a complex living form). Tellers of such tales derive their plots from the fact that mutations are indeed random events. But the plots ignore the important point that natural selection "sifts through" individual variations created by the mutations. In other words, natural selection non-randomly retains those traits that happen to be adaptive from a reproductive standpoint. Random shuttling of bricks will, in fact, eventually build a castle if only those accidental arrangements corresponding to intact walls are allowed to remain. Obviously, it would take an enormous amount of time to build a castle in this way, but time is not a severe constraint in evolution (having taken place over the age of the earth, 4.5 billion years). Another simple analogy, described in the next paragraph, will drive the point home.

Suppose I draw 10 cards face-down from a deck. The odds of the cards being all-red are very small. If a set of 10 cards is not all-red, then I place all the cards back in the deck, shuffle, and try again. It would take a prodigious amount of time (unless I got very lucky) before a randomly selected set of 10 cards would produce 10 red cards. Now let us change the rules a bit. Again, I select 10 cards. But this time any red that I happened to pick is retained, while the blacks are placed back into the deck. Assume, for example, that the first 10 cards were 6 blacks and 4 reds. I keep the reds (they have "survival value"), return the 6 blacks, and deal myself 6 new cards. Of these, 3 are black and 3 are red. I add the 3 new reds to the 4 that I already have, giving me now 7 reds. Since I need 3 more to reach 10, I randomly deal myself 3 cards, two of which happened, in this particular example, to be red. I now have a total of 9 reds. Another random draw turns out to be red, and I have therefore reached my goal of 10 reds. Thus, in only 4 deals I have quickly achieved 10 red cards. Of course, I have exercised a form of "natural selection" by retaining the "beneficial" cards (the reds). In summary, by imposing a selection process upon the random distribution of cards, I have reduced the time it takes to achieve a "successful species" (the 10 reds) from days (or weeks) to a few minutes.

The lesson from the simplistic card game is clear: Natural selection is not a pure "adaptation by chance" mechanism. It is not a tornado in a junkyard making an airplane. It is not a blind watchmaker making a clock. Evolution is produced by random mutations working hand-in-hand with a preserving force called "selection".

One cannot ask too much of random mutational changes because any major structural modification in an organism is almost certain to be fatal. The more complex an organism, the more susceptible it is to mutational damage. A random gunshot at a radio, for example, is likely to do more serious structural damage than a random gunshot at a simple object such as a hammer. Thus, a slight improvement resulting from a beneficial mutation, occurring every once in a while among far more prevalent destructive mutations, is all that can be hoped for. Human beings will never suddenly develop a third eye or sprout angel wings. Evolution takes place only by means of small variations on existing structures over extended time periods. Darwin expounded clearly on the principle of "descent with slow and slight modification":

> That natural selection generally acts with extreme slowness I fully admit.

> As natural selection acts solely by accumulating slight, successive, favorable variations, it can produce no great or sudden modifications; it can act only by short and slow steps.

> It may metamorphically be said that natural selection is daily and hourly scrutinizing, throughout the world, the slightest variations; rejecting those that are bad; preserving and adding up all that are good; silently and insensibly working whenever and wherever opportunity affords.

Suppose an organism weighed 10 kilograms (22 pounds). Suppose further that there is a favorable selection for larger body weight such that the average body weight increases only 0.01% per generation. Thus, the change in the first generation would be virtually unnoticeable, increasing from 10 kilograms to 10.001 kilograms. But if this selection continued uniformly for 1000 generations (a short time on the evolutionary time-scale), the average weight would increase to 27 kilogram (almost 60 pounds). This example illustrates the rather obvious point that small variations, taken over huge periods of time, can produce major structural changes.

Darwinism may be considered a rather crude theory in that it lacks quantitative detail. "Slow and slight modification" was never defined either in terms of number of years or the magnitude of structural change. Since in the 1850s Darwin had no clear idea as to the age of the earth, it is perhaps a good thing that he did not attempt to assign specific time-values to his theory. Even today, however, descriptors of evolution as "slow", "rapid", and "explosive" are bandied about without definition. Intense debates have appeared in the recent literature over whether traits appeared gradually or suddenly, but the corresponding time-values are often not mentioned. The problem is compounded by the fact that "time" means different things to different disciplines. What might be slow to a geneticist could be fast to an archeologist. Fortunately, for the purpose of this book, it makes little difference whether an organism appeared on the scene one million years or, for example, four million years ago. Although certain scholars would seemingly commit homicide in defense of one date or the other, I often regard a factor of 4 as trivial. I do not mean to belittle those who gather, usually with great effort, evidence for a particular time-value. I merely state that my description of neo-Darwinism, and my ultimate modification of it, do not depend upon a few-fold uncertainty in time.

Darwin claimed that evolution proceeds by gradual, nearly imperceptible steps. A variation on this theme, punctuationism, is worthy of special mention owing to recent attention to the subject. Consider, by way of illustration, the case of horse teeth. Over a period of 60 million years, horse molars have evolved from small branch-eating teeth to large grass-grazing teeth. The diameter thereby increased by 9 mm for an average of 0.15 mm per million of years. By all counts, one would have to regard this as a gradual change in line with Darwinian thinking. But there are those, particularly Eldredge and Gould, who would argue that the change was not necessarily gradual. There could have been long periods of no change ("stasis") interspersed by periods of rapid changes. The combination is called "punctuationism". Thus, horse molars might have enlarged in a punctuated manner: 2.0 mm the first 10 million years, not at all during the second, 4.0 mm during the third, and 1.0 mm for each of the three remaining 10 million years (for a total of 9 mm in 60 million years).

Although some feel that punctuationism is revolutionarily anti-Darwin, Darwin would, no doubt, have had little problem accommodating the concept. It all depends upon one's particular time-frame. For example, as mentioned above, over a 60 million time-frame the horse molar evolution may indeed be considered as punctuated. Over the third 10-million-year period, molar evolution takes place faster (0.40 mm/million years instead of 0.15 mm/million years). Assume now that this rate of 0.40 mm/million years occured uniformly over the entire 10-million-year period. Therefore, when contemplating only that particular 10-million-year period, one would be led to think that molar evolution is gradual. The conclusion here is that one's viewpoint on gradualism depends upon the time period under scrutiny. Thus, tooth evolution could have been simultaneously punctuated over 60 millions years but gradual over 10 million years.

Punctuationism is not a theory that violates the essence of gradualism. It simply postulates (quite reasonably and totally consistent with the fossil record) that changes are usually not uniform throughout a long time period. Since puctuationists can be viewed as tremulous gradualists, the idea of punctuationism is a mere wrinkle in the history of evolutionary thought. Two points are, however, agreed on by nearly everyone: (a) An almost imperceptible gradualism, postulated by Darwin, is not always consistent with the fossil record. (b) At the other extreme, the appearance of major structural modifications (as in the formation of a new species) in just a few generations (called "saltation") has never been observed and is not a viable concept.

There are those who claim that gradualism, punctuationism, and saltation are all distinctly different. There are those, like myself, who prefer to regard punctuationism as a variant of gradualism. There are even those who equate punctuationism with saltation. The confusion stems, as mentioned above, from evolutionists who describe the duration of events using descriptors that are never defined. For example, one source calls punctuationism "brief episodes of rapid gradual change". This lack of rigor has less to do with carelessness than with the absence of information. Consider horse teeth again. Searching for old horse teeth encased in hard rock is brutal work. Empty hands among the few hardy fossil-seekers must be a common experience. As a result, insufficient horse teeth are available to plot a complete time-course for tooth development. We do not know if horse tooth evolution was gradual-smooth or gradual-intermittent. Nature has a secret past, and she is reluctant to reveal all of it even with arduous prodding.

If lack of information is a frequent problem, the complexity of Nature is an ever-occurring one as illustrated by the famous case of cats and clover. In "*The Origins*" Darwin speculated that the welfare of certain flowers in English villages are affected by the number of cats. How can this be? The answer goes something like this:

The cats keep down the mice known to destroy humble-bee nests. Humble-bees are the exclusive pollinator of red clover. Thus, we have the following sequence (including one additional component suggested by a later writer): Old maids tend to own cats; cats eat the mice; with fewer mice around, humble-bee nests are less ravaged; the resulting population increase of humble-bees then pollinate the clover. Clearly, old maids (unwittingly of course) favor the survival of clover! Darwin used this story (minus the old maids) to demonstrate the complex interrelationships in Nature and the difficulty of discovering the precise way in which the struggle for existence operates. He also made the important assertion that competition could be most severe among the closely related forms:

> As the species of the same genus usually have, though by no means invariably, much similarity in habits and constitution, and always in structure, the struggle will generally be more severe between them, if they come into competition with each other, than between the species of distinct genera. We see this in the recent extension over parts of the United States of one species of swallow having caused the decrease of another species.

Similar species occupying similar environmental niches meet head-on in competition for food, shelter, and other resources.

Variants best able to cope with competition, whether the competition be intra-species or inter-species, pass on their traits preferentially to their descendants. Thus do organisms change over time. These statements sound reasonable, and yet they are beset with an unpleasant and even unscientific vagueness owing to the fact that the term "competition" is ill-defined and non-quantifiable. The words "cooperation", "progress", "diversity", and "struggle" have similar difficulties, in contrast to terms such as "gene", "phylum", "carnivore", "bipedal", "polygamy", etc. that have much more specific definitions. Since a certain level of fuzzy terminology used by modern evolutionists is unavoidable, one should not be discouraged if, on occasion, it leads to something less than a full clarity of understanding.

The basic principle of evolution has several corollaries (all of them saying essentially the same thing): Evolution does not anticipate. Evolution makes no room for long-range considerations. Evolution has no purpose in view. Evolution does not occur to produce a beneficial effect upon an organism. Evolution does not operate per se for the good of the species. Evolution does not proceed by following the canons of optimal design. Evolution is indifferent to the harmony of the ecosystem.

The preceding assertions derive from the fact that mutational and other genetic variations are, as already explained, random. It is thus impossible to predict how a particular species will evolve. An eagle's eye has not been developed because some ancient bird ancestor "foresaw" the need for an eagle's eye. An eagle's eye has developed because the sight of its ancient ancestors gradually changed (each tiny step representing an enhanced probability of surviving and reproducing) until today there is a bird, called an eagle, that owns a magnificently evolved eye. The story is one of success but not one of deferred success.

Incidentally, there is no logical reason to suppose that any two particular features of an organism must evolve simultaneously in one smooth, coordinated program. No doubt the development of various organelles and organs were, in general, out of phase. *Archaeopteryx*, the primitive bird, had feathered wings but reptilian teeth; birds did not lose their teeth until 80 million years later. Evolution of feathered wings and tooth-loss were obviously independent events.

Eighty-one species of frogs have been identified in one square mile of Equadorian jungle. The diversity of similar animals in a small ecosystem has been used to argue against the plausibility of natural selection. Why should Nature, in its slow step-by-step progression, fill an environmental niche with so many related but different animals? My answer to this question could be one that I, as a student, always hated to receive from my professors: "Why not?" I see no particular reason why there must be a single best-suited and dominant frog species. There need not be only one set of biological features that survives in a given ecological niche (not that a square mile of jungle is necessarily a single niche). A particular frog species might be better in one respect (catching flies) while another species is better in a different way (more eggs produced). Over the eons, a delicate balance among species is reached (accounting for the disastrous effects often observed when man disturbs the environment). To summarize: Since genetic variations are random, modifications can radiate in a host of successful directions.

Along a similar vein, one can point out many examples of a single species, or type of animal, living under amazingly diverse conditions. I personally have been bitten by mosquitoes all the way from the arctic to the tropics. A small burrowing wasp inhabits the steamy forests, the dry deserts, and the moderate climes of Australia. Eucalyptus trees grow all over the world, wet or dry, warm or cool, sea level or upland. This remarkable adaptability need not cast doubt on natural selection theory. All it means is that a given set of structural features can survive and reproduce successfully in more than one ecological niche. Human beings originally evolved, it is believed, in Africa. Ultimately, humans migrated to other areas, including the arctic, where they continued to do well. In contrast, there is a species of grasshopper that lives only in Stone Mountain, Georgia near my home. The grasshopper does not know this, but of the two criteria of evolutionary "success" (population and geographic diversity), it is less prosperous than the humans and mosquitoes that blanket the earth. Many, however, consider this grasshopper a valuable citizen nonetheless, and they are saddened whenever such specialized creatures leave the scene forever which is happening far too often.

Perhaps, like me, the reader has wondered about an aspect of natural selection that, at least at first thought, appears a bit troubling. To illustrate the problem, consider a group of foxes and rabbits. Within the rabbit group an individual appears on the scene that can run slightly faster than the others. Over time, the descendants of this rabbit will occupy an increasing fraction of the population because the variant can better outrun its enemy, the fox. This is all very good for the rabbit, but not so good for the fox.

Fortunately for the fox, however, a fast-running variant also appears among this predator. Having a more reliable rabbit diet, and thus a higher reproductive capacity, the "fast fox" expands its presence in the population. But would this not continue back and forth among rabbit and fox until a supersonic fox is chasing a supersonic rabbit? Where is the logical flaw in the conclusion?

One answer is that there are physical limitations as to how fast a particular animal form can run. The limitations are imposed by leg-length, metabolism, lung and heart capacity, etc. If it happened that evolution reached a stage where a prey can run faster than the speed attainable by a predator, then the predator might indeed become extinct. This is no doubt why Nature seems to be in balance; disrupted equilibria, achieved over the eons, tend to be eliminated.

Wolf and moose have been shown to coexist in a more-or-less equilibrium. The wolf avoids healthy moose, with a potentially lethal kick, while culling the weak and infirm from the herd. Often the latter are moose with worn-out teeth that do not allow the animal to eat properly. The wolf thereby receives the food it needs, and the moose herd benefits from a positive evolutionary force in which the strong survive and reproduce more frequently without competition for food from weak members of the herd. Wolves are not super-predators, nor are all moose unfailingly successful in escape.

Another reason for the absence of super-organs or super-capabilities is their cost to the animal. It takes energy, for example, to construct an improved bone and muscle system for running. If running faster means having fewer pups, or producing less milk, or growing thinner fur, then the advantage of running faster comes into question. Life is basically a trade-off. And it is natural selection that determines exactly how the traits are balanced among each other.

The trade-off principle is well illustrated by the soil nematode, *C. elegans*. These worms are self-fertilizing hermaphrodites that produce, at maturation, about 300 sperms and a much larger number of oocytes (eggs). Since nearly every sperm is used to fertilize an egg, the maximum fecundity (offspring production) is around 300 per worm. Now a mutation was discovered that produced 500 sperms instead of 300. At first sight, this mutation would seem to be of the type that is favored by natural selection. But such is not the case. Even though the mutant had a higher fecundity, the population of the original "wild-type" worms grew more quickly under standard conditions. How can this be? The answer is that the mutant required 2.5 hours longer to mature all its sperm than did the non-mutant wild-type. It turned out that this delayed sperm production was more than sufficient to eliminate the greater than 50% advantage in sperm output. Greater sperm numbers in the mutant was associated with too high a price.

One should not get the impression that natural selection is a stalwart march toward perfection. All organisms have traits that are less than ideal; they are not what you or I would have designed had we been given the opportunity. Few women, I presume, would praise their unique estrous cycle system. Humans have so frequent problems with their back that it is difficult to believe that a better musculature is not possible. In fact, no less than 120 human features have been listed that could warrant improvement (and I can personally vouch for many of them). There are several reasons for imperfection in Nature. An organism may, for example, not yet have had time to adapt fully to an environmental change. When human ancestors left the forests, they began to walk on two limbs ("bipedalism") from which the species no doubt experienced an overall benefit. Perhaps they could better spot enemies in tall grass, or more easily carry game back to camp, or better avoid the tropical sun. We pay the cost for this change in walking style, however, with a less than perfect back which is prone to give us pain and trouble.

Probably the most common reason for a lack of perfection in Nature relates to the fact that traits do not necessarily evolve independently from one another (although, as we saw in the case of feathered wings and toothlessness of birds, they can do so). Living structure, it must be remembered, is locked into an extremely complicated interdependence, and what ultimately evolves is a compromise. Imagine a mutation that affects ten different traits (not an unlikely possibility). Assume that eight trait-changes are adaptive, whereas two of them are out of synchrony with environmental needs. Naturally, it would have been preferable to the species to have a mutation that gave ten improvements, but the developmental and structural constraints of the organism might not permit such a mutation. Alternatively, a ten-improvement mutation, although in principle possible, might simply never have occurred in the random modification of the genetic material. The important point here is that the "8 good/2 bad" mutation will persist in the population as long as the net change, taken over the entire organism, is favorable to survival.

It is easy to construct a fictional animal with an awesome potential for survival: fangs of a cobra; scent glands of a skunk; quills of a porcupine; claws of a cat; breeding capacity of a rabbit. Since these accoutrements would require a great deal of energy to produce and maintain, I would make the animal both carnivorous and capable of eating almost any plant. Fortunately for us, Nature working with random variations, never discovered the genetic formula for this nightmare. As Darwin stated: "Natural selection tends only to make each organic being as perfect as, or slightly more perfect than, the other inhabitants of the same country with which it comes into competition." Owing to this truism, human beings can, by modifying an environment, quickly render an organism slightly less than perfect for dealing with its competition, and the organism disappears.

Natural selection is usually regarded as a mechanism for change, but it can also serve as a mechanism for constancy ("stasis"). If an organism is well adapted to a niche (or series of niches), and if most generations of reproductive adults equal or exceed in number that of the proceeding generation, then the organism might be under little pressure to change. Natural selection would effectively remove the variants as they appeared. Shark, horseshoe crab, dragonfly, and gingko tree are all examples of organisms that have survived relatively unchanged for millions of years. Along similar lines, a particular trait, such as a five-digit hand, may become permanently fixed within the genetic framework. Apparently, five digits worked well in the distant past, and switching to, say, six digits cost too much for too little additional benefit (if there were any benefit at all).

A biological trait may have originally evolved for a reason other than its present function. For example, some evolutionists believe (without hard evidence) that bird feathers initially developed for reasons of warmth rather than flight. In this manner, the formation of primitive feathers, which have little flight value, can be explained. Cases are also known where organs adapted to one set of conditions may persist, and therefore seem out of place, when new and different environmental conditions have rendered them useless. Darwin himself described the example of plants that have hooked seeds beautifully designed for transport on the fur and wool of animals. Plants with hooked seeds have found their way (perhaps via flotation) to oceanic islands devoid of mammalian life. If, over time, these island plants became modified into a new variety or species, but retained the original seed hooks, then it would appear to the casual observer that evolution has created a useless trait. Retaining a useless trait, such as the appendix in humans, is likeliest when the trait costs the species little energy to produce and maintain.

Perhaps it is time to pause and take stock. Why am I discussing these various components and examples of natural selection? I am doing so because it is necessary to give the reader a firm grasp of the elements of Darwin's great theory. In order to understand the deficiency of the theory, one must first appreciate its strengths. Too many books have been already written that take a painfully biased position (both positive and negative) on Charles Darwin. "*The Thin Bone Vault*" is, I hope, not another of these.

Thus far I have used the word "mutation" a great deal, but, as already mentioned, mutations are not the sole source of genetic variation. Genetic drift is one of many other factors that might also play a role. Genetic drift is evident in small, localized populations that do not communicate with each other. For example, wolves in geographically distinct populations will differ slightly in coat color, length of limbs, size of ears or tail, etc. In general, no two separate local populations are ever exactly alike even in the absence of strong pressure from natural selection. Genetic drift takes place because the odds are high against identical changes occurring in two physically isolated groups.

Darwin was enthralled with animal breeding, devoting the first pages of "*The Origin*" to raising of pigeons. Centuries ago, humans began breeding wolves or wolf-like animals for desirable traits. Selection for speed gave greyhounds; for aggressiveness, pit bulls; for sense of smell, blood hounds; for pack instinct, the shepherd dog. Special mutations may have been exploited on occasion, but for the most part breeders depended upon "natural" variations in the dogs' genes. (No two members of a species are genetically identical unless they are twins). Differences among dogs today are astounding. A German shepherd does not even look the same species as a Chihuahua. Yet if they could physically manage the act, they would mate and reproduce. As different as dogs appear, however, they are clearly "dogs". No one would ever confuse cat-sized dog with a cat. Breeders have succeeded only in making "canine cartoons" by exaggerating inherent qualities. These human-imposed exaggerations seem to have reached a limit. Since the initial supply of genetic variability has more-or-less been used up, a rat-sized dog is not imminent. In contrast to dogs, cats seem to have maintained much of their original wild nature. Efforts at breeding have been less intense and, more important perhaps, their genetic material is less variable and more resistant to tampering by man. Each species has its own characteristic susceptibility to change via recombination of genes through sexual reproduction. The genetic variability in man is an interesting question that will be addressed later in this book.

The preceding paragraph implies that dog-breed diversity emerged largely from traits inherently present in the animal's genetic makeup. According to some, however, genetic mutations are a likelier source of variability. Although mutation rates of DNA are normally very low (too low to explain dog-breed evolution in a short time span), in certain sections of dog DNA (called "tandem-repeat sequences") mutations are 100 000 times faster than normal. Mutations in these sections might have given rise to the altered appearances and breed diversity. Darwin himself, of course, had no access to genetic concepts such as mutation and considered breeding as the sole selective force.

As Darwin did recognize, animal breeding has produced new varieties but no new species. This did not bother him. In fact, he rarely discussed the origin of species despite the title of his book. Darwin simply regarded species-creation as another example of "descent with modification". A group of animals isolated on one side of a mountain range would change so much that it eventually could no longer mate with its former compatriots on the other side of the mountain. A new species was born. This obviously cannot be the whole story, but the scenario is considered by many as a good approximation of how new species originated.

When pressed to cite specific examples of current-day evolution, textbooks frequently (and erroneously according to recent claims) call attention to "industrial melanism" in a species of British moths. Until 1845, the moth was known only in its "peppered" form that had dark markings on white wings. In that year, a totally dark form was discovered in the industrial and sooty city of Manchester. Fifty years later, the dark form of the moth increased from 1% to 99% of the population. In smog-free areas of England, the original light colored variety persisted. Natural selection is the obvious explanation. The dark moth, resting on soot-covered trees and buildings, was more difficult to be detected by predacious birds. Since dark moths had a survival advantage over the peppered ones, the former soon became the predominant variety. Other examples of observed natural selection include the recent formation of penicillin-resistant bacteria. Those mutant bacteria that have acquired an ability to destroy antibiotics survive preferentially when exposed to the drugs, and they soon dominate the bacteria population.

One might think that the above examples "prove" natural selection. Unfortunately, things are not quite this simple. Moth coloration and bacterial drug resistance are categorized as "microevolution" because the changes are small (requiring the action of only a single new mutation and corresponding enzyme). Most people, including many creationists, agree that natural selection controls microevolution. "Macroevolution" (called "real evolution" by some), involving gross structural development and the creation of new species, is another matter. Here the mechanism for change is under serious debate as we saw in the quotes from scientists at the beginning of *Chapter 1.*

Note that my discussion has emphasized physical traits and said little about behavior. Darwin realized the importance of behavior but made no serious attempt to integrate it into his theory. He did write, however, that "the most wonderful of all known instincts, that of the bee-hive, can be explained by natural selection having taken advantage of numerous, successive, slight modifications of simpler instincts." Today we know that many behavioral patterns of the bee and other insects arise from organic compounds, called "pheromones", whose production is genetically controlled. For example, a dead ant emits a "funeral pheromone" that compels its living sisters to carry the deceased to a burial pile many feet from the nest. A stinging bee deposits an "alarm pheromone" that causes other bees to get angry and attack the same spot (which is why one can be chased through the woods by a group of bees). Non-instinctive behavior in human beings is the subject of considerable controversy, particularly with regard to how much of it is inherited (and thus subject to classical natural selection) and how much of it is learned or cultural. There exists no proof, for example, that political fervor has a genetic basis connected to natural selection. On the other hand, an inherited hormone imbalance could conceivably lead to abnormal aggression and even to a predisposition to anti-social behavior.

A spectacular example of a "wired in" behavior pattern is displayed by a fly known as *Hilaria*. The male flies manufacture balloons of silk and offer them to a particular female. The female accepts the balloon she likes best, and agrees to mate with the fly that made it. How did this ritual evolve? No one has the slightest notion, of course, but it is known that there are male flies of another species that present a female something more substantial (a paralyzed grub) in order to win her favors. Other varieties of flies wrap their grub in silk prior to giving it to the female. Apparently, *Hilaria* males discovered that the female could be fooled by offering her the silk packaging without any gift inside! And somehow or the other, this trickery became fixed into the genetic makeup of the flies where it is essential to the survival of the species.

As a final point, I might mention an effect given the fancy name of "*adaptive phenotypic plasticity*". All this means is that an organism can occasionally alter its physiology rather than its genetics in order to accommodate an environmental change. For example, adult animals of identical genetic composition might all become generally smaller in times of food shortage. To some extent, such adaptive effects render unnecessary the inherited trends that natural selection would otherwise establish. "Adaptive phenotypic plasticity" does not negate natural selection but may in certain instances delay it.

Thus ends our brief encounter with natural selection. The purpose of this chapter was to capture the spirit of Darwinism in a reasonably objective and concise manner. The greatness of the natural selection idea lies in its ability to explain Nature with a magnificent simplicity. When one stops to think about it, natural selection seems to make intuitive sense, and one wonders why it took as long as it did to finally develop the concept.

Mankind was both illuminated and humbled by Darwin. Biologists suddenly saw themselves as a part of Nature, rather that separate from it, and the world was never the same again. Even if natural selection were suddenly shown to be completely false, "*The Origin*" would still remain one of our greatest intellectual achievements. In the next chapter, I take up many of the objections that have been levied against natural selection by scientists and non-scientists alike. I will defend Darwin against the criticisms because they are, in my opinion, misleading, specious, or outright false. Moreover, my addressing the criticisms will be instructive in that it will lead to a more thorough understanding of natural selection. Ultimately, as it turns out, we do indeed finally encounter situations that force an expansion of the natural selection doctrine. This comes in the second part of the book where I delve into the "thin bone vault" that houses human intelligence.

Chapter 3

DARWIN ANALYZED

In this chapter, I will consider widely cited criticisms of natural selection as a mechanism for evolution. Criticism within evolutionist circles should not be regarded as unhealthy or as a sign of impending doom for Darwinists. Quite the contrary. Debate over natural selection is symptomatic of the discipline's vigor — of the continued attempt by the thoughtful to reach the full truth. There are, of course, those who have exploited differences of opinion among scientists to further a religious cause. This nuisance merits two comments. First, under no circumstances should the search for new understanding be impaired by fear that a tiny minority will abuse the information. Second, many claims by creationists (e.g. that the earth is 6000 rather than the actual 4.5 billion years old) are articles of faith and thus irrelevant to this particular book. But other concerns of creationists (e.g. the gaps in the fossil record) are also concerns of evolutionists and, to this extent, must not be ignored. In summary, the discussion in this chapter is limited to issues raised by serious scholars. Although creationists may have absorbed many of these issues into their own rhetoric, this is incidental to the purposes of this book.

Part 1: The Evolutionary "Story"

It may seem strange, but one of the most common criticisms of natural selection is that it explains too much. This is partly the result of the unchanneled fervor of evolutionists who, in their admiration of natural selection, seemingly couch all of Nature in its terms. The prize for such silliness can be given to a well-known naturalist who suggested that roseate spoonbill evolved its bright pink coloration so as to be less visible to its enemies at sunset. Such rationalizations fill the literature, and I call them *"evolutionary stories"*. And this is exactly what they are: speculative "stories" in the adaptive mode. I cite below examples of stories to emphasize the need for evolutionists to exercise great care when attributing "purpose" to each and every trait or behavioral nuance encountered in Nature.

Human Intelligence. Darwin himself, in his "*Descent of Man*", fell victim to story-telling when he attempted to explain why human males are, ostensibly, more intelligent than human females: "The greater intellectual vigor and power of invention in man is probably due to natural selection, combined with the inherited effect of habit, for the most able men will have succeeded best in defending and providing for themselves and for their wives and offspring". (Darwin must be forgiven both for his sexism and for his support for inheritance of acquired habits; after all his work is over a century old and must be judged in that context).

Darwin was saying (if I might paraphrase) that men are more intelligent than women because, in the course of their hunts, the men had to exercise stealth and cunning. Those best endowed with the mental capacity to catch game had the highest chance of survival and reproduction. Since women did little hunting, they would not be exposed to such selective pressure. The point here, however, is that I could just as easily have invoked natural selection to rationalize why women are the more intelligent. Women stayed at camp and cared for the young, an activity that required the development of language and communication skills. Women may also have had to solve technical problems such as building fires with wet wood, tanning hides, and fashioning pottery. Thus, evolution of intelligence favored the female sex. Who is correct, Darwin or myself?

The answer is that neither of us is correct; our evolutionary stories are equally unprovable and meaningless.

Moles. The mole (*Talpa europea*), as well as certain cave-dwelling fish, has lost the ability to see. Eyes of such animals have either atrophied or disappeared altogether, and skin often covers the useless organs. One can comprehend the survival value of vision, but why should vision, once in place, be discarded in animals when they move into an ecological niche where light never penetrates? An evolutionist has suggested that lack of eyes among the inhabitants of the dark provides a survival advantage in that the visual organs are no longer around to become infected. Infected eyeballs are postulated as an evolutionary force! Destruction of a trait is explained as casually as its appearance.

Critics of Darwinian theory, in response to stories such as that of the blind mole, often complain that natural selection "explains everything and therefore nothing". What exactly does this mean? The comments refer to the fact that a powerful theory will usually have predictive power. In the world of chemistry, for example, one can predict the structure of hydrogen using the theory of quantum mechanics. The theory of natural selection also has predictive power, but it is a capability necessarily limited by its embodying an element of chance. Let us attempt to predict, for example, whether a fish living in a dimly lit cave will possess eyes. The fish might have eyes so as to allow catching prey a trifle easier. On the other hand, the fish might not have eyes owing (ostensibly) to the danger of fatal eye infection. Clearly, a prediction is not possible because it is unclear which effect will dominate within a given species in a given environment. Evolutionary stories that glibly explain all eventualities, including both the presence and absence of a trait, serve only to emphasize the inherent limitations of natural selection. Stated in another way: a theory thoughtlessly expanded to cover everything cannot predict anything.

Female Orgasm. Unlike male ejaculation, female sexual satisfaction is not necessary for reproduction. Thus, the female orgasm might seem puzzling from an evolutionary point of view. It has been suggested that female orgasm serves the female in helping to select a male partner who is patient and attentive to her needs rather than being rough and impulsive. Such a male is, according to the story, more likely to be a patient and caring father who will look after a child during the many years required for the child to reach adulthood. This would contribute to the child's survival and, thereby, to the continued propagation of the parents' genes. Doubt as to the validity of this sentimental story has recently arisen from experiments showing that a female orgasm might in fact promote the intake of sperm and thereby assist fertilization and continuation of the species.

The "good lover/good father" story is at least harmless, but Darwinian-based arguments can also lead to unsupportable and highly mischievous conclusions. A good example is seen in a quote from a well-known scientist, E.O. Wilson:

> In hunter-gatherer societies, men hunt and women stay at home. This strong bias persists in most agricultural and industrial societies [*sic*] and, on that ground alone, appears to have a genetic origin...My own guess is that the genetic bias is intense enough to cause a substantial division of labor even in the most free and most egalitarian of future societies...Even with identical education and equal access to all professions, men are likely to continue to play a disproportionate role in political life, business, and science.

Accordingly, women are regarded as being poorly endowed genetically for participation in the professions. Although Darwin may be forgiven for his sexism, such abuse of natural selection in modern times is less excusable. The ease with which evolutionary stories can be constructed to support a prejudice is all the more reason to exercise caution in their use.

Cicadas. There are three cicada species each of which has a 13-year and a 17-year variety. Every 13 or 17 years (depending upon the variety), the insect will erupt from the ground in spectacular numbers. There is, apparently, something special about the numbers 13 and 17 because three different species of cicada utilize these time intervals in their reproductive cycles. It has been proposed that values of 13 and 17 years have evolved because they are prime numbers. A prime number is a number that cannot be divided by any smaller number except 1. Thus, 3, 7, and 11 are prime numbers but 4 (divisible by 2) and 9 (divisible by 3) are not. If the cicada emerged in a non-prime number of years (say, every 9 years), then the insect would be susceptible to parasites possessing 3-year cycles. But a 13- or 17-year cicada cycle makes it much more difficult for a parasite to synchronize its life cycle with that of its host. Of course, the rationale does not explain why, among all the possible prime numbers, only 13 and 17 were selected. Nonetheless, the idea is certainly a clever one, and — who knows — it may contain an element of truth.

Homosexuals. Critics of natural selection are fond of citing homosexuals who, it is argued, should have disappeared from the population long ago owing to their low reproductive rate. Darwinism requires, of course, preferential reproductive capacity in order to incorporate a trait permanently. Defenders of natural selection have responded with the suggestion that in prehistoric times, homosexuals, being free from family responsibilities, were available to help their relatives with hunting and other activities essential for life. The homosexual trait was thus favored and preserved indirectly via the progeny of heterosexual relatives who possessed overlapping genomes with the homosexual family member. This "sociobiological" rationale is deceptively appealing, but, in actual fact, there exists no hard information as to the origin of homosexuality. It is just another example of an unproven (and unprovable) evolutionary story that has been widely publicized in texts, journals, and university courses.

Cuckoos. Cuckoos lay their eggs in the nests of redstart warblers and allow the redstarts to do the caretaking. Even though the cuckoo nestlings may kill the smaller redstart babies, the redstart parents continue to feed the cuckoos. A diehard natural selection advocate has concluded that the situation must be best for both birds. If the behavior were not advantageous for the redstart, so the argument goes, then the redstart would have evolved a defense mechanism against the parasitism. Perhaps there is an overabundance of redstarts, and the loss of a few nestlings is harmless or even overall beneficial to the species. This is a possibility, but one can just as easily postulate that the redstart neurology is too firmly set to allow the development of a defensive response (e.g. pushing the cuckoo nestling out of the nest). Or, perhaps, the cuckoo behavior has evolved too recently for the redstart to have had time to respond genetically. Many evolutionary stories are possible, and one is as good as another. We can appreciate the comment of Francis Crick (of DNA fame): "The trouble with evolutionary biologists is that you are always asking 'why' before you understand 'how'."

It may appear that my comments on evolutionary stories are censuring Darwinism rather than defending it. This is not meant to be the case. I merely wish to protect Darwinism against certain overly enthusiastic evolutionists who, without evidence, ascribe all of Nature to natural selection and, in so doing, unnecessarily expose the theory to skepticism or even disbelief.

Sloths. It is simply not possible to find clear-cut "reasons" for all the features, even important ones, of animals and plants. As already described, this has not stopped people from trying. Consider the tree-dwelling sloth that defecates at intervals of greater than a week. When the animal does decide that the time has come to relieve itself, the sloth descends to the ground (where it otherwise seldom goes) for the express purpose of burying the feces it had dropped from the sky. Why does the slow-moving sloth expose itself to danger on the ground for little apparent gain? The evolutionary "explanation" is as follows: In the course of evolution, a series of random mutations created an ancestral sloth that had a propensity to bury its feces. It so happened that this behavioral trait was beneficial to the particular tree species in which the sloth often lived and fed. As the tree thrived with the fertilization, so also did the sloth that provided the fertilizer, and the burial trait persisted. Whether it is true or not, one cannot be immune to the charm of such evolutionary stories.

Giraffes. In 1827, a giraffe was put on display at the Museum of Natural History in Paris. It was noted how wonderfully adapted the animal was to reach vegetation high up in the trees. At that time many people believed in Lamarck's theory that acquired traits can be inherited. Thus, the giraffe would "stretch" itself while feeding on the highest leaves and, over many generations, the animal permanently obtained a long neck. American school children are taught to regard Lamarck's giraffe as the epitome of a stupid idea. Actually, Lamarck never discussed giraffes in his original writings. It was Darwin, decades later, who brought up the subject of giraffe necks: "The giraffe, by its lofty stature, much elongated neck, forelegs, head and tongue, has its whole frame beautifully adapted for browsing on the higher branches of trees. It can thus obtain food beyond the reach of other Ungalata or hoofed animals inhabiting the same country and this must be of a great advantage to it." Darwin, however, was also missing the mark. The neck of the female giraffe is two feet shorter than the neck of a male giraffe. If a long neck were critical to survival, a famine would likely cause the females to starve in preference to the males (a highly counter-evolutionary effect). In actual fact, giraffes spend far more time grazing than they do cropping trees. The "reason" that their necks are long is probably that their legs are long, and they need to reach the ground to graze and drink. Thus, even the classical Darwinian story, the origin of long giraffe necks, is suspect. It teaches us, however, an important lesson: A trait may exist not because of some independent function (e.g. eating tree branches) but because another trait (e.g. long legs) requires the first trait's presence. This is all the more reason to exercise great care when concocting possible responses to the "why" of a trait.

Growing Old. Many scientists in the past believed that senescence in human beings is a genetically programmed event whereby

Nature purposely removes the elderly. In this manner, the drain of resources on the young (who are engaged in the all-important reproductive activities) is mitigated. Taken as a whole, the species benefits. By way of anecdotal support, we have all read how Eskimos place their elderly on ice floes when they become too infirm. Actually, the Darwinian explanation of senescence is, in all likelihood, just another example of specious story-telling. After all, for most of human evolution our life expectancy was closer to 30 than the current 75. Although our ancestors no doubt had lethal encounters with blizzards, disease, saber tooth tigers, and each other, they were spared the modern experience of senescence. Genes for aging probably arose because natural selection provided no good mechanism for their removal. If genes kick in after our reproductive period has passed, then those genes are less subject to selective pressures, and Nature may be indifferent as to whether the genes are beneficial, harmful, or neutral. Thus, senescence may be a trait without a "purpose"; it may simply exist in our genetic makeup.

Gazelles. When a predator threatens a herd of Thomson gazelles, one or two gazelles in the vicinity of the predator may begin to leap ostentatiously in high arcs. This so-called "strotting" behavior has been explained in terms of an individual selflessly calling attention to itself in an effort to save others in the herd. There are, however, serious reasons to doubt this story. One would imagine, first of all, that predation would quickly eliminate a sacrificial strotting trait from the population. Second, it is unclear how strotting preserves the species because one or another gazelle will likely get eaten in any event. Overall survival value would be achieved only if a predator ate a single strotting gazelle in preference to two or more non-strotters (hardly very likely). According to an alternative selectionist argument, the strotter is attempting to save itself as opposed to its compatriots. In effect, the strotting gazelle is telling the predator: "Hey, you ought to go after someone less agile and easier to catch than myself." This anthropomorphic story has its problems too. If there were in fact survival value to the "selfish" behavior, then why has it not spread throughout the herd? Note that there is at least one virtue to these gazelle stories that distinguishes them from many evolutionary tales: The two alternatives (i.e. selfless and selfish behavior) can be differentiated by experimental observation. One merely has to record whether strotting gazelles get eaten more frequently (with selfless gazelles) or less frequently (with selfish gazelles) than the others in the herd. I am unaware of such studies having been carried out.

Razorbills. Razorbills are big, black and white seabirds that nest in colonies on cliff ledges by the sea. At one time it was believed that razorbills mate faithfully with only a single partner, but recent observations say differently. Apparently, just prior to laying her eggs, the female razorbill will often slip away from her mate to another ledge where she copulates with a second male. Even after the eggs are laid, the female tends to visit another ledge for a brief encounter while her mate is back home incubating the eggs. Since these "extramarital affairs" occur when the female is infertile, one might wonder as to how they became ingrained in the razorbill behavioral genes. A current hypothesis points out that the bird colony returns to the same cliff year after year. Accordingly, the females take advantage of any opportunity to "audition" males in order to find the best prospect for next year's mating. Hence, the genes for adulterous behavior have been preserved. Through suitable tagging of the birds, it should be possible to test whether this year's lover is, indeed, next year's mate. One could by this means differentiate the hypothesis from my proposed alternative "cultural" possibility: The female just likes to have a little fun on the side; and happy birds are reproductively successful birds.

Butterflies. There is a species of butterfly that possesses "protective coloration" (i.e. the insect blends into the surroundings to make its detection by predators more difficult). This same butterfly does a peculiar thing as a post-fertile adult: it drops to the ground and beats its wing rapidly until it dies of exhaustion — a form of suicide. How could self-obliteration possibly benefit the species? According to the prevalent theory, the post-fertile adults become a burden to species survival by providing clues to predatory birds — clues as to how to discriminate the butterfly from its surroundings. When non-productive adults remove themselves from the population, they are no longer around to "teach" predators the limitations of the camouflage. This is no doubt an ingenious idea; as an exercise, the reader might try to devise two other stories that are equally probable. When the exercise is completed, the reader should reflect on the considerable difference between conjecture (stories that might or might not be true) and stories that actually encompass experimentally proven information.

I will now end this section with three final stories involving human beings.

Human Societies. Sociology, political science, and economics have not escaped armchair theorizing and storytelling in the adaptive mode. A good example of "social Darwinism" can be found in the 1911 edition of the *Encyclopedia Britannica*:

> The first organized societies must have been developed, like any other advantage, under the sternest conditions of natural selection. In the flux and change of life, the members of these groups of men which in favorable conditions first showed any tendency to social organization became possessed of a great advantage over their fellows, and these societies grew up simply because they possessed elements of strength which led to the disappearance before them of other groups of men with which they came into competition. Such societies continued to flourish, until they in turn had to give way before other associations of men of higher social efficiency. In the social process at this stage all the customs, habits, institutions and beliefs contributing to produce a higher organic efficiency of society would be naturally selected, developed, and perpetuated.

At one time in the United States the above "survival of the fittest" philosophy provided a convenient "explanation" for the dominance of the "advanced" white race as it invaded the territory of the American Indian. Regrettably, Darwinism is such a broadly based theory that it can be, and has been, exploited to justify warfare, the culling of "undesirables" from society, and many other horrors. Darwin himself wrote, "care for the imbecile, the maimed, and the sick is...highly injurious to the race." He expressed the hope that such people would refrain from marriage.

Generosity to Strangers. There is no direct evolutionary benefit to ourselves from being good to a person who is not kin to us (i.e. a person with no genetic overlap). Yet examples of humans even risking their lives to save strangers are commonplace. Sociobiologists have developed a theory of "reciprocal altruism" to account for such kindness to strangers. The idea is that person B, who has been done a favor by person A, will eventually return that favor to person A or his family. By this indirect path, the probability of survival and reproduction of person A will have been enhanced. Some theorists go so far as claiming that civilization would not have been developed without this ingrained reciprocal interchange mechanism. In fact, some have even proposed that intelligence evolved to enable a person to differentiate a "cheat" among a group of otherwise fair-minded reciprocators. I, frankly, am skeptical about the "I perpetuate your genes, you perpetuate mine" concept. There are so many stark cases of people placidly watching the demise of a stranger, rather than rescuing him or her, that I wonder about the generality of the "kindness to strangers" premise. In place of reciprocal altruism, I might suggest the possibility that we are kind to strangers (on occasion, that is!) when inherent generosity to our own kin cannot be completely "turned off" in the face of a non-family member in need. Thus, kindness to strangers is merely a spin-off or by-product of a genetically self-serving tendency to help our own. Of course, one also cannot exclude the possibility that saving a stranger from drowning has more to do with personality traits like courage, desire for admiration, and enjoyment of excitement than it does with reciprocal altruism.

Menstruation. "Every mechanism out there was designed by natural selection to solve a problem, so you have to identify the problem", declared evolutionary biologist Margie Profet as she was wondering about possible explanations for menstruation. One night in 1988, according to her account, she had a dream about menstruation (a dream that she remembered because her cat woke her up in the middle of it). The dream revealed that menstruation is valuable because it allows the reproductive tract to rid itself of pathogens that happen to attach themselves to sperm. Not so, claims biologist Beverly Strassmann; there is no evidence that there are more pathogens in the uterus before menstruation than immediately afterwards. Instead, she proposes that the uterine lining sloughs off because the alternative, keeping the womb in a constant state of readiness, would require more energy than does the menstruation cycle and its periodic renewal. Later in the book, energy costs will be shown to be a critical component in the evolution of intelligence.

In summary, evolutionary stories attempting to relate "why" certain features have evolved are frequently speculative and unprovable. If evolutionary stories are so unsatisfying, then why should I be defending Darwin? The fact remains that one cannot deny the validity of natural selection simply because certain of its proponents abuse, or at least overuse, the concept. Natural selection constitutes a viable theory for the evolution of many plant and animal traits despite its inability to serve as a blanket explanation for all (or even most) features of life. This inadequacy stems not from the falsehood of the theory so much as from the complexity and randomness of life, as well as from the almost unfathomable amount of time that was required by life to evolve relative to our own pitiful lifetime. Natural selection may ultimately require modification, but this must be done on the basis of sound scientific principles. Misuse of a theory by certain of its proponents does not constitute grounds for ridicule or dismissal of the theory itself.

One final point with regard to evolutionary stories: As we have just seen, many stories answer the question "Why?" in a highly superficial manner. To understand the distinction between superficial and not-so-superficial, consider the following question: "Why does the male gypsy moth mate with the female gypsy moth?"

A superficial reply would be: "Mating behavior evolved to propagate the species; without it, the species would become extinct." A reply with a greater information-content would be: "Mating behavior occurs because, as a result of past mutational events, the female produces a sex pheromone whose odor induces copulatory behavior in the male." Both answers are correct, but the quality of the answers differs. The superficial response emphasizes the function of mating rather than the mechanism of its occurrence; little is learned from it. The alternative answer, while being incomplete, incorporates more definitively the cause of the behavior (i.e. the factors controlling it). Likewise, when an evolutionist states that "a bird has a thick beak in order to crush seeds," one is informed more of the function of the beak than how it actually came into existence. Once the tenets of natural selection are covered, I plan to focus much more intensely on the question of "How?" than on the question of "Why?" as applied to the human brain. I presume we all have a good idea of why the human brain has evolved its capabilities; the question of how the capabilities came to be is far more interesting.

Part 2: Darwinism as a Scientific Theory

A scientific theory inevitably leads to well-defined consequences and expectations. Any theory will be discarded if the expectations it engenders are not realized in the course of future research. "Predictiveness" is, therefore, one of the most important elements of a scientific idea. Gravitational theory, for example, predicts the acceleration of a rock dropped from a height before the rock is ever dropped. Einstein's famous $E = mc^2$ allows the energy E to be calculated from a mass m and the speed of light c without actually making the measurement. This is not to imply that experimentation and observation are unnecessary. Experiments and observations are needed to stimulate the production of theories and then, later, to confirm them. The key point here is that a theory, substantiated by experiments and observations, is inseparately linked to a remarkable and powerful attribute: Predictiveness.

It has been asserted that natural selection is not a valid scientific theory because the theory cannot predict the course of evolution. No one has, for example, applied the principles of natural selection to "predict" the fact that reptiles are capable of evolving into birds. No one can "predict" what the pigeon will look like 1000 years from now. Do such uncertainties disqualify natural selection as a useful scientific idea? Absolutely not. Natural selection necessarily fails to predict the future because it embodies within it the element of chance. As already described, evolution depends upon random genetic changes that are retained according to their ability to impart reproductive advantage to the organism. It is simply not possible to foresee the consequences of random genetic change upon the evolutionary development of an animal or plant.

An inability to foretell what particular genes will mutate (and what the physiological consequences will be once the mutations occur) is not the only contributor to uncertainty. Scientists are unable to forecast long-range changes in the environment (the environment, of course, determining the reproductive advantage or disadvantage of the mutations). How can one predict that, in the future, an animal's fur will become thicker when it is not known whether the climate will get colder? The most one can say is that if the climate gets colder, then natural selection will probably favor many animals acquiring warmer coats. But this is a simple, almost trivial, microevolutionary example. The minute a really interesting case is considered, such as the evolution of a totally new species, then problems in prediction become insurmountable. We remain ignorant of (a) what genetic changes and gene interactions are required to create a new species; (b) the probability that mutations, taking place randomly, will in fact materialize in the proper sequence; and (c) how environmental circumstances will interact with the organism to favor or disfavor its survival. In summary, the predictiveness of natural selection with respect to species evolution is currently non-existent and probably will always be so.

No law of Nature says that humans are inherently capable of understanding everything. (If this comes as a surprise, consider the likelihood that we will ever know what it is like to be an armadillo!)

Scientists, of course, do not often admit to their limitations. Like most everyone else, scientists prefer to extol their successes rather than confess to their innate inadequacies. Yet the limitations of human understanding cannot be ignored especially with regard to biological features that required millions of years to evolve. Certain aspects of evolution are, of course, poorly understood because historical facts (e.g. fossils) are hard to come by. I am not referring to this informational component of our ignorance. I am, instead, expressing doubt that we, as recent descendants of primitive cave-dwellers, are smart enough to understand complex biological characteristics no matter how much evidence we are likely to collect. We may not, for example, be sufficiently intelligent to comprehend (in anything but a superficial manner) this same intelligence.

Given these problems, especially at the species level, why is natural selection nonetheless a valid and useful theory? One answer is that natural selection can indeed be used to make certain "predictions". Listed below are a few of them:

a) Bones of a primitive primate will appear in older rock than bones of a more highly advanced primate.
b) If a given species of animal is larger now than it was a million years ago, then specimens who lived within the intervening time period will have an intermediate size.
c) No grazing animal is likely ever to have "counter-evolutionary canine teeth".
d) A colony of mixed black and white hares introduced into an arctic region will, over time, become entirely white.
e) An animal that roams the forest during the night will have evolved with acute eyesight.
f) It will be possible to trace the geneology of a given animal type living in a confined area, such as an archipelago, where an ancestral species has diversified into descendant species. This step-by-step evolutionary change within a genus (i.e. a group related species) will be demonstrable via their morphology, gene sequences, and enzyme patterns.

Clearly, the predictions are not very "risky". Nor are they all hard-and-fast. For example, a grazing animal might have vestigial canine teeth in the unlikely (but conceivable) event that the animal evolved from a carnivore. Or the prediction of intermediate body weights of animals living in intervening time slots could be false if evolution did not involve "smooth" transitions. In addition to a measure of uncertainty, the predictions are also non-quantitative and lack the breadth common to predictions in the physical sciences (e.g. Einstein's $E = mc^2$ where m can be *any* mass of *any* element). Nonetheless, natural selection provides a framework within which we can intelligently contemplate Nature and, at least to a limited extent, foresee what might happen in the future.

My prediction that white Arctic hares would eventually replace the black ones in snowy regions of the Arctic is another case in point. Although invisibility to predators is of obvious survival value, one can never be certain which trait (or combination of traits) will ultimately determine viability. If "blackness" were linked to, say, greater warmth via heat absorption, then black hares might have a survival advantage over white hares, visibility to predators notwithstanding. The fact that Arctic hares are now all white in winter might have more to do with natural selection never having "discovered" black genes than with white being inherently more viable than black. My prediction that white hares would win out was, I confess, more of an educated guess based on how natural selection has already operated not only on the Arctic hare but also on species such as the arctic fox and the polar bear. I am reasonably certain (but not completely certain) that natural selection would move in the same direction after my introducing hares of mixed color. That is to say, I am reasonably certain (but not completely certain) that Nature has already optimized the winter coloration of the arctic hare. My uncertainty is, in part, derived from the penguin which, although living happily in the Arctic, is half black. And whereas the ptarmigan changes to white in the winter, the spruce grouse, living side-by-side with the ptarmigan, remains brown all year.

There is still another problem with the "black-to-white" arctic hare prediction apart from its lack of certainty. If, on testing, my prediction is verified, it would be a "success" only at the *microevolutionary* level (i.e. at the level describing rather superficial and easily attained modifications). Microevolution, as with industrial melanism in moths mentioned earlier, is rigorously documented, and few people have any quarrel with it. Extrapolating and magnifying microevolutionary events to the creation of complex organs, not to mention entirely new species, is another matter. Predictions are, it is safe to say, usually beyond reach within this *macroevolutionary* domain.

We see, therefore, that the physical sciences obey strictly deterministic laws, while evolutionary biology obeys (in the unkind words of J. Hershel) "the laws of higgledy-piggledy". This disparity between the physical sciences and evolutionary biology must never be attributed to an inherent falsehood of natural selection. Instead, the difference is directly related to the fact that responses to selective pressures are probabilistic, thereby precluding the possibility of making in-depth predictions. One must not expect that natural selection provide more than it is prepared to give and then, being disappointed, cast natural selection into the realm of a pseudoscience. Thus, philosopher of science Karl Popper was unfair when he once wrote: "Darwinism does not really predict the evolution of variety. It therefore cannot really explain it." One might more accurately state: Darwinism explains the evolution of variety, but cannot really predict it; the explanation is, therefore, very useful but less powerful than those of the physical sciences.

One reason that scientists have retained an affinity for natural selection, despite its inadequacies, is that it is the only rational theory around. Reject natural selection, and there remains no scientific understanding at all. It is possible (as believed by many great physicists of the past including Bohr, Schröedinger, Pauli, and Delbrück) that there are important laws of biology that have yet to be elucidated. Poorly understood laws of biological self-organization and embryology are among these. Until such laws are in hand, natural selection will, no doubt, continue to be invoked in explaining how organisms are interrelated in a process of ancestry and descent.

I may add, as a historical note, that scientists frequently retain a theory, even though it encounters anomalous data, until such time that a better theory emerges. For centuries scientists defended the Ptolemaic theory of the heavens in which the earth served as the center of the planetary system. Puzzling observations, such as the trajectory of Mars, were either ignored or explained away with complicated and unbelievable "epicyclic" motions. Ptolemaic theory was ultimately discarded not because of its difficulties but because Copernicus devised a better theory in which all planets encircle the sun. In retrospect, one might ask: "Why was it not obvious to astronomers that the Ptolemaic theory was fraught with problems and incorrect?" The answer is simple. The problems were tolerated until a simpler, more powerful, and more predictive theory came along. Similarly, problems with natural selection will be tolerated until an improved construct is developed. When there is only one theory available, those who wish to discard it inevitably face a burden of replacement.

In concluding this section, note that evolutionary biology is a historical science (like geology) as opposed to an experimental science (like chemistry). Thus, evolutionary biology depends upon observation, intuition, deduction, and an appeal to everyday experience in order to devise the most economical explanation for the origins of life. Experimentation is difficult because one cannot condense in a few hours or days what took Nature millions of years to accomplish. Moreover, no experiment can ever provide proof of a past event in the first place. I might, for example, be able to document the assertion that last year my cat had a litter of kittens (using dated photographs, witnesses, etc.), but I could hardly devise an experiment to test the assertion. Likewise, no experiment will shed light on the reproductive advantages of a species in the Pleistocene era. William Jennings Bryan showed he did not understand these simple concepts when he said: "I will believe in evolution when I can sit in my garden and see an onion turn into a lily." Bryan demanded the impossible: experimental proof of a species-to-species conversion within the human time-frame.

Part 3: Reductionism

Science in the twentieth century has been dominated by a highly successful approach called "reductionism". This philosophy comes into play when a system is too complicated to study in its entirety. Consequently, the system is subdivided ("reduced") into small portions in hopes that, eventually, information from the small portions can be combined to construct the system as a whole. For example, it may be difficult to study a particular chemical reaction within a living cell where countless other processes are going on simultaneously. One may, therefore, address this difficulty by examining the reaction as it proceeds, all by itself, in a test tube. In this manner definitive information will be more easily obtained (although the question of the relevance of the information to the total cellular system is ever-present).

In order to contrive certain important features of reductionism, I will use the concept of a prime number. As mentioned earlier, a prime number is a number that is divisible only by one or by itself. Thus, 3, 5, and 7 are prime numbers, whereas 9 is not because 9 can be divided by 3. All even numbers are, of course, non-prime because they are divisible by 2. Now let us ask, and answer, questions about three sets of numbers as the sets become smaller and smaller (more "reduced"):

a) How many odd numbers between 0 and 20 are *non-prime*?
   Answer: Two
   Note: Only 9 and 15 are non-prime; all other odd numbers in the set are prime.
b) How many odd numbers between 0 and 14 are *non-prime*?
   Answer: One
   Note: Only 9 is odd and non-prime in this set.
c) How many odd numbers between 0 and 8 are *non-prime*?
   Answer: None

Notice that when the largest set (0 to 20) is reduced to the smallest set (0 to 8), the answer becomes more definitive. Stating that there are two odd non-prime numbers between 0 and 20 imparts a measure of uncertainty because it is not clear (without further analysis) exactly which two odd numbers are non-prime. In contrast, the answer for the smallest set (0 to 8) is "none". The answer "none" reveals something precise about each member of the set, namely that they are all either even or prime. One sees that as the set is reduced, a given piece of information "tells us more about less". This is reductionism in action.

The numerical example also teaches another lesson: Since the answers for all three sets are different, one cannot extrapolate properties of the most reduced set (0 to 8) to the two larger sets (0 to 14 and 0 to 20). Herein lies the danger of reductionism. What is true for a component alone is not necessarily true for a larger system incorporating that component. Extrapolations going up the complexity ladder are risky.

Now let us apply some of these ideas to an example from evolution. Ultra-Darwinist Richard Dawkins wrote the following:

> The total amount of suffering per year in the natural world is beyond all decent contemplation. During the minute that it takes me to compose this sentence, thousands of animals are being eaten alive, many others are running for their lives, whimpering with fear, others are being slowly devoured from within by rasping parasites, thousands of all kinds are dying of starvation, thirst and disease. It must be so. If there is ever a time of plenty, this very fact will automatically lead to an increase in population until the natural state of starvation and misery is restored.

Few have ever described so dramatically the dismal life of the non-humanoid animal world. The question then becomes: Can we extrapolate the situation from this reduced system (i.e. only non-humanoid animals) to the whole system (i.e. all animals including humans)? As we have just seen, extrapolations are a risky business. Risky or not, Dawkins claims that the experiences of the non-humanoid animal subset can indeed be extended to humans (with the clear implication that human evolution has faced the same environmental pressures as other animals):

> In a universe of electrons and selfish genes, blind physical forces and genetic replication, some people are going to get hurt, other

> people are going to get lucky, and you won't find any rhyme or reason in it, nor any justice. The universe that we observe has precisely the properties we should expect if there is, at bottom, no design, no purpose, no evil, no good, nothing but pitiless indifference. DNA neither cares nor knows. DNA just is. And we dance to its music.

Note the extreme reductionism where it is asserted that in a universe of mere electrons there can be no good, no evil, no justice. In other words, electrons and DNA do not care about human welfare, and therefore humanity is doomed to "pitiless indifference" from Nature. Confusion here derives from a classical reductionist error, namely an unwarranted extrapolation. Although both animals and humans are composed of electrons and DNA, humans are not ordinary animals. Humans are endowed with a unique organ housed in the "thin bone vault". This organ, the brain, is the seat of intelligence, the mind, the conscious, self-awareness, call it what you will. Were it not for our special brain, we would indeed be constantly running for our lives, suffering from rasping parasites, even beating our fellow competitors to death. Actually, this is not quite true. Were it not for the brain, serving as a counterweight to our animalistic tendencies, we would have become extinct long ago. Humans are, after all, without great speed, without claws, without warm fur, without night vision, without sensitive noses, without lethal teeth, without venom. Fortunately, our unique mental capacity has saved us.

One may legitimately ask why a text sympathetic to Darwinism takes issue with an ultra-Darwinist. The problem is that certain Darwinists, in their exuberance, do a disservice to the field of evolutionary biology by making assertions that few can believe. Who can accept (or want to!) the argument that our genes doom us to live in a dog-eat-dog world never moderated by morality, civility, generosity, and good sense to name a few?

I hasten to add a final point. The fact that one cannot always extrapolate from animal to human should in no way imply superiority of human over animal. The quote I like best in this regard comes from nature writer Henry Beston:

> We patronize the animals for their incompleteness, for their tragic fate of having taken form so far below ourselves. And therein we err, and greatly err. For the animal shall not be measured by man. In a world older and more complete than ours, they move finished and complete, gifted with extensions of the senses we have lost or never attained, living by voices we shall never hear. They are not brethren, they are not underlings; they are other Nations caught with ourselves in the net of life and time.

Part 4: The Wonder of It All

Although many observations can be rationalized by Darwinism, natural history also presents a large array of phenomena that do not lend themselves to any in-depth explanation. Several books, even recent ones, explicitly cite these phenomena as evidence that Darwinism is false or at least suspect. The argument is in essence: "This biological feature is amazing. How could natural selection ever create such a thing?" My position is a little different. I share in the wonder of the accomplishments and ingenuity of Nature, but I do not elevate that wonder to a "disproof" of Darwinism. I can marvel at the mysteries of Nature without expecting natural selection to explain it all. The unfathomable features of our living world should spur further research rather than serve as a destructive force with regard to what little understanding we do possess. Cited below are puzzling items from natural history that defy easy explanation.

Electricity-Producing Animals. Electric eels (*Electrophorus electricus*) can produce one amp at 500 volts, sufficient to light ten 50-watt bulbs. Yet the eel, mysteriously, does not electrocute itself. Similarly, certain species of shark and bony fish have perfected an extraordinarily complex "electrolocation" system in which electric impulses are used to sense objects in murky waters. No one has a clue as to how such elaborate electricity-generating organs, and their supporting structures, managed to evolve, or why such a formidable capability (with its seemingly high survival value) has not prevailed more widely throughout the animal kingdom.

Gastric Brooding Frogs. The gastric brooding frog (*Rheobatrachus silus*), extinct since 1981, swallowed its eggs and then brooded about twenty tadpoles for 6–7 weeks in its stomach. Ultimately, the mother "gave birth" by burping up her froglets. If the species evolved from a more conventional type of frog, as seems likely, then the following changes would have had to take place:

a) The mother frog had to learn to swallow its eggs rather than deposit them in the water as is now done by all other species of frogs.
b) The mother's stomach chemistry had to be radically altered so as not to kill the young with her digestive enzymes and strong stomach acids.
c) Passage of the eggs from the stomach into the intestines had to be suppressed.
d) The tadpoles had to be converted from a mobile, feeding organism into one that could survive when imprisoned for weeks in a dark, crowded frog stomach.
e) The burping event had to be programmed so as not to occur too early nor too late in the tadpole development (either of which could have been fatal).

One has no problem in positing an evolutionary advantage to stomach brooding: Direct development from egg to tadpole can thereby occur without the young being subjected to predators and other dangers of the external world. A puzzle exists, however, in the fact that each of the above steps required substantial behavioral or physiological modifications in the brooding frog's presumed ancestors. Since failure at any point would have been disastrous to the frog's survival, one must conclude that several evolutionary changes manifested themselves in a coordinated fashion. In other words, several chance mutations and other genetic changes (and no one has any idea how many) assembled at the correct time and place, by a mechanism too complicated and too remote to understand.

Actually, there is one aspect of *Rheobatrachus* chemistry that has been well worked out. The frog eggs have been shown to produce a substance called prostoglandin-E2. This compound inhibits the production of gastric acid in the mother's stomach and, in so doing, helps convert the stomach into a brooding chamber. The eggs must have fortuitously evolved prostoglandin-E2 for some unknown purpose prior to the first successful brooding event. So when a mother frog fortuitously ate the eggs (presumably for food), the eggs were fortuitously able to suppress their own destruction by acid in a stomach that fortuitously could "switch-off" its digestive enzymes. Fortuitously, the mother burped her brood at exactly the right time before the young died from over-crowding or the mother died from starvation or a burst stomach.

"Preposterous", says the critic of evolution. "These fortuitous events, all occurring in concert, are too unlikely to believe". One can sympathize with the sentiment; the brooding frog is indeed an amazing creation of Nature. But before dismissing natural selection on this basis, we must recall that at one time in the earth's history the brooding frog did not exist. But there is no denying that despite this improbability, the frog obviously did indeed make its appearance on the scene. How exactly did this happen? We do not know, but it is reasonable to presume that natural selection played a role. Gastric brooding must have given the frog a reproductive advantage. And never lose sight of the fact that evolution had a long time to occur — too long for the human mind to adequately grasp. This may not be a very satisfying response to "How exactly did this happen?", but at present it is all we have.

Cellulose Digestion. Humans have evolved hundreds of enzymes (i.e. proteins that catalyze biochemical reactions). We possess enzymes that digest fats, sugars, and other proteins. How strange it is that we lack an enzyme to digest cellulose, the major component of wood. It is not as if cellulose were a particularly exotic substance. In fact, cellulose is merely a string of sugar molecules and similar in structure to starch, a substance we do digest easily. Bacteria in the gut of termites and cattle produce an enzyme that splits cellulose into its component sugar molecules, but we humans have never developed a gene that codes for this particular enzyme. If we had acquired a "cellulose gene", then we could obtain our calories simply by eating wood or hay (which would have been a valuable trait especially in times of starvation). The lack of such a simple and obviously beneficial gene seems all the more puzzling when one considers the presence of eyes, ears, kidneys, etc. that require not a single gene but an extremely complex set of genetic modifications. If humans can have such complicated organs, why can't we have a simple digestive enzyme as generated by a single gene? There is no answer to this question except to say again that Nature does not necessarily work toward perfection. To survive extinction, a species need only be "good enough" to maintain its population, and humans certainly do not seem to have any problems on that particular score.

This section, entitled "The Wonder of It All", has cited examples of the electric eel and brooding frog, whose evolutionary bases defy simple explanation. Now let me approach the issue from a slightly different angle. I will list items that, like the lack of a cellulose enzyme in humans, seem counter-evolutionary. That is to say, the items seem (at least outwardly) to be harmful to the best interests of an organism. I will then attempt to explain, in a totally ad hoc manner, possible reasons for their presence. This exercise is useful to rebut widespread claims that seemingly counter-evolutionary traits invalidate the theory of natural selection.

a) Honey bees have their stingers connected directly to their abdomen. When stinging an animal, the bee deposits its stinger in the animal's skin, a process that fatally tears out the bee's abdomen. The honey bee need not necessarily have evolved such a suicidal mechanism as seen from the yellow-jacket, which can engage in multiple stinging with no harm to itself.

Does the honey bee stinger not run afoul of evolutionary expectations?

Response: Perhaps the honey bee, in the random modifications occurring over the course of its evolution, never acquired a genome that allows for a stinger to "disconnect" from the intestines as possessed by the yellow-jacket. The required mutations simply never occurred, and the honey bee apparently did fine without them.

b) The rhinoceros has, typically, only one calf every four years. This seems like an unnecessarily low reproductive rate with which to propagate the species. Why would such an inefficient system evolve?

Response: No one can deny that if the rhinoceros had a calf every year, or even every two years, there would be more rhinoceroses in the world. Unfortunately, the low reproductive rate of the rhinoceros, coupled with poaching by humans, might ultimately result in its extinction. Had the rhinoceros evolved a more efficient means of expanding its numbers, (requiring, it would seem, only an accelerated estrous cycle), the danger might not exist. This does not mean that natural selection has "failed" and that, therefore, the theory is a failure. Chance mutations and selection processes led to a species that, at least in the past millennia, successfully maintained its population. Conditions have changed with the appearance of humans, who shoot the animal for its horn, and under such circumstances the reproductive capacity may no longer be sufficient to maintain the species. Evolution is a slow process and cannot always keep pace with changing conditions. Nature, of course, does not "care" whether there are a multitude, a few, or none at all. It is only we humans, with our ability to feel emotional pain, who recoil from the idea of species becoming extinct or from the idea of a drab overpopulated world where mammals are limited to ourselves and our domestic animals.

c) Human males grow hair on their faces even though, one would think, a beard would allow easier grabbing by enemies. It is possible that facial hair provided warmth, but then one must wonder why human females, who need warmth also, have not been similarly endowed.

Response: Facial hair in human males is a direct consequence of testosterone production. Testosterone production was, no doubt, useful for survival since the earliest evolutionary times when hunting, fighting, and other dangerous activities were daily experiences. The male beard may have been retained as a disadvantageous by-product of testosterone. If this is true, then clearly the beard was not sufficiently disadvantageous to outweigh the overall benefits of the hormone. The lesson here is that one cannot properly consider a trait's advantage or disadvantage in isolation from all the other traits that are genetically linked to it. Although such complexity greatly hinders the predictive utility of natural selection, the theory cannot be dismissed on the basis of superficially disadvantageous traits.

d) Many butterflies imitate the complex coloration of a different butterfly species that is distasteful to predators ("*protective mimicry*"). Yet few butterflies are green despite the fact that green is an easy color to synthesize in Nature and that a green coloration would render a butterfly nearly invisible among the foliage. How is it that Nature missed out on such an obvious survival mechanism?

Response: It is true that a green coloration would make insects more difficult for a predator to detect among the foliage (a fact utilized by caterpillars but, to my knowledge, not by most butterflies). Perhaps the beautiful markings of a butterfly have another purpose inconsistent with a green color. One is tempted to speculate that the markings serve as a sexual attractant except for the fact that sexual attraction among butterflies is mainly chemical ("pheromonal") rather than visual. A male gypsy moth, for example, can detect the sex-attractant pheromone of a "calling" female up to six miles away. Perhaps, however, at some time in the past butterflies did indeed locate their mate via visual cues, but over time, a more efficient and long-range pheromonal system evolved. Thus, butterfly markings may be an evolutionary relic with no purpose at all (except perhaps to delight the human onlooker).

e) The first-hatched of many birds will kill their younger siblings even in periods of abundant food. This seems illogical for a species fighting for survival.
Response: Many first-born nestlings are genetically programmed to kill their younger nest-mates by pecking them or shoving them out of the nest. Perhaps the genes for this behavioral pattern evolved during a long period of food shortage when it was better for one young bird to survive than for all to die. Apparently, no additional mechanism has ever evolved (advantageous though it might be) to "turn off" the fratricide even when the food supply is plentiful. Thus, under favorable conditions, the trait appears illogical; its true value is apparent only under adverse conditions.

f) Millions of American chestnut trees have been killed by an imported fungus. An attempt was made to create a resistant mutant of the American chestnut by X-ray irradiation. It was hoped that such a mutant would be attainable because a closely related species, the Chinese chestnut, is completely resistant to this same fungus. Thus, the necessary genetic alterations in the American chestnut should not be too difficult to achieve. Yet no resistant mutant was ever obtained even though X-rays are known to greatly accelerate mutation rates. If random mutations are such an important evolutionary force for survival, why is it that one cannot obtain a beneficial mutant in this relatively straightforward evolution-mimicking experiment?
Response: Gross, uncontrolled molecular damage of DNA, as caused by X-ray irradiation, is most often harmful or outright lethal. Only rarely does something useful come of it. Evolution has had sufficient time to wait for a beneficial mutational event, buried among a vast number of deleterious ones. Mankind, on the other hand, has but a handful of years with which to experiment. Thus, the experiment with the American chestnut could, in principle, work; its failure is related mainly to a necessarily short time-frame. There is, incidentally, an alternative

approach to solving the chestnut problem. One could identify the gene(s) in the Chinese chestnut responsible for fungus resistance. This gene(s) could be implanted into the American chestnut genome via typical genetic engineering methods (discussed later in the book). But this is a genetic manipulation by design; *random* mutations in Nature require extensive time periods to achieve desirable genetic alterations.

g) Blood-drinking bats are found in South America but not Africa. Sea-snakes common in the Pacific and Indian oceans are absent in the Atlantic. Is it not logical that natural selection should have, instead, populated such similar niches with similar animals?
Response: First, the so-called "similar" niches may not be as "similar" as perceived. Subtle, sometimes unnoticed differences may play a key role in the forms of life that ultimately evolve in them. Second, even if two niches were identical in all respects, there can be independent evolutionary pathways leading to distinct species in each of them. This is, of course, less likely if the two niches are physically connected, but even here the situation is not clear. Independent evolution might still occur when the rate of communication between the two niches is slow relative to the rate of evolution. Moreover, it is possible that a geographical interconnection, now in place, may not have existed in the past when evolutionary paths were set into motion.

The list of biological phenomena that are ostensibly "illogical" from an evolutionary standpoint could be extended to the point of tedium. No doubt all scientists, even the most ardent evolutionists, admit to being puzzled by many products of Nature. The question is, however, not whether Nature can be puzzling but whether, as some have claimed, the "illogical" traits discredit the theory of natural selection. As just seen, I have given many examples where the question is answered negatively. Admittedly, in certain instances (e.g. with bird fratricide being understandable under conditions of poor food supply) I used an "evolutionary story" to make my case. Although I have previously recommended caution when inventing "evolutionary stories", I do on occasion ignore my own advice. I may have absolutely no evidence for a speculative response to a seemingly illogical trait, but a conjecture nonetheless makes a worthwhile point, namely those "illogical" traits may not be as illogical as they first appear.

In 1991, an article appeared entitled "Biologist Discovers that Survival of the Common Orchid Challenges Darwin's Natural Selection Theory". The article mentions that the lady's slipper orchid "thrives across the Eastern United States and is one of the most common naturally occurring orchids in the country". For 14 years, biologist D. E. Gill examined 3300 individual flowers and discovered that only 23 of these had been pollinated. Reproduction for the orchid is difficult because the flower produces no nectar to attract its only pollinator, the bumble bee. Instead, the orchid depends upon deception (i.e., it resembles a nectar-producing flower also visited by the bumble bee). But bumble bees are no fools. After their first visit to a lady's slipper, they quickly learn to avoid the flower and, consequently, seldom attempt to visit an orchid a second time as would be necessary to achieve cross-pollination. It was speculated that the 23 pollinations that did occur happened by an "accidental second visit" or by a bee "with a high level of desperation". Lack of nectar was definitely a source of the orchid's problems because when nectar was artificially placed onto the lady's slipper, the pollination success markedly increased. Another curious fact is that the lady's slipper could, in principle, with only a rather minor flower modification, pollinate itself (as do many other orchid species) and thus avoid the need for bumble bees altogether. The investigators could "see no reason why either of these two alternatives — nectar production or self-pollination — would not be the winning things to do". Hence the challenge to Darwin's theory.

After reading this article, one might be tempted to exclaim: "This orchid is maladjusted. Survival of the fittest (if I may use this phrase) would never have produced such a pathetic lack of fitness." I almost fell into this trap myself when I happened to re-read the beginning of the report mentioning that the lady's slipper orchid "thrives across the Eastern United States and is one of the most common naturally occurring orchids in the country". The conclusion seems obvious. Lady's slippers do fine (at least when undisturbed by humans) despite a reproductive system that, to us, appears badly designed. Perhaps biologists have not yet uncovered all the secrets of lady's slipper reproduction; perhaps they have. Whichever the case, one can hardly invoke an obviously successful flower as an argument against natural selection.

Birds have not developed (as have some butterflies) a chemical that makes them distasteful to predators — with at least one notable exception, the rubbish bird. Biologists J. Dumbacher and B. Beehler captured a rubbish bird while netting birds for tagging in New Guinea. In attempting to free the bird from the net, the scientists received minor scratches from the bird's sharp claws. In a rather instinctive response, the scientists licked their wounds only to feel a hot, badly tasting, lingering sensation on their tongues. This unpleasant taste was traced to a highly toxic chemical exuded by the bird. The chemical was identified and found to be (strangely enough) the identical compound as in the skin of an Amazonian frog from which the local Indians poison their arrows. A single rubbish bird can produce enough of the toxin in its skin to kill a good-sized hawk by paralyzing its nervous system. Of course, hawks are not stupid enough to eat the bird once they get a taste of it; they quickly release their prey and, presumably, do not bother the species again. The rubbish bird has, clearly, evolved an effective survival ploy.

The poison of the rubbish bird cannot be all that difficult to "biosynthesize". As already been pointed out, in the Amazon, far away from New Guinea, a frog has independently developed the same defense mechanism. One might argue that natural selection could have easily imparted such protection to all birds subject to predation, and the fact that this has not happened casts doubt on the theory. In actuality, the absence of a common toxin system among birds favors, rather than discredits, natural selection. After all, selection processes cannot come into play until such time that a variation appears. And the main source of variation, mutation, is a matter of pure chance. I can no more disclose "why" among all the birds only the rubbish bird has evolved the poison than I can explain "why" an eight-of-hearts was drawn out of a deck of cards.

Everybody knows that parrots can say phrases such as "Polly wants a cracker". Suppose someone inquired as to how on earth the parrot evolved an ability to enunciate an English sentence. One might, in response, invent one of those "evolutionary stories" as follows: Through random mutations, the parrot's nervous and vocal systems developed an ability to make a wide variety of sounds including those of other species. With this ability "in hand", the parrot can attract these other species into its locality, thereby diverting the attention of predators off itself and onto the newly arrived birds. To complete the story, one would have to add something to the effect that the nervous and vocal systems capable of imitating birdsongs carries over to the elaboration of sounds in human languages. No one should take my contrived story seriously. There is a far better response to the question, "Why has natural selection favored genetic variants of parrots capable of saying 'Polly wants a cracker'?" The answer is, "It has not....English did not even exist when the parrot evolved". It might not be very satisfying to state that a parrot mimics English sentences "because it can", but at least one thereby avoids pretentious stories to explain the wonders of nature.

In summary, the wonders of Nature, observed almost on a daily basis, in no way disprove natural selection. Some of the puzzles can be explained with stories that are, perhaps, plausible if not provable. Other mysteries reflect the fact that natural selection is limited in scope and should not be expected to explain everything in Nature of which we are curious about.

Although I have defended natural selection against unfair criticism based on the wondrous creatures of Nature, I must also defend the wondrous creatures of Nature from indiscriminate use of natural selection. It is one thing to teach that the earth's flora and fauna are the products of natural selection; it is quite another to teach that the earth's flora and fauna are *merely* the products of natural selection. Ah, how a single word makes a difference!

For one thing, we know too little to claim that natural selection is the sole source of biodiversity; we do not even understand how something as basic as the cell membrane came into being. More importantly, however, the use of the dismissive word "merely" is harmful to the human spirit in that it contributes to our losing a hallmark of our species: A highly developed sense of wonder. As we dissect Nature into its historical components, we must always stand in awe of the spectacle. Anthropologist M. Konner expressed the same sentiment:

> It is up to us to try to experience a sense of wonder about the earth that will save it before it is too late. If we cannot, we may do the final damage in our lifetime. If we can, we may change the course of history and, consequently, the course of evolution, setting the human lineage on a path toward a new evolutionary plateau.

An earth that does not elicit our sense of wonder is an earth that is not respected and an earth that is, therefore, abused. And, as Konner suggests, an abused earth could well be our undoing.

Part 5: Entropy

Entropy! Only the "uncertainty principle" ranks above it in the degree of public abuse endured by an important concept in physics. I recall once watching a painter in a Chicago park demonstrating his skill on a warm summer's day. The artist heaved small cups of paint onto a canvas to produce, as one might imagine, a colorful but indescribable mess. He turned and said to me: "My painting is a manifestation of the uncertainty principle". Similarly, a friend once told me that he was not going to clean up his backyard because it was a hopeless battle against entropy. Entropy (a particularly difficult concept to understand in detail) has nonetheless entered the folklore to become, in a corrupted form, a source of confusion in many areas of popular discussion, including evolution.

Think of entropy as equivalent to "disorder". An increase in entropy means an increase in disorder (or decrease in order).

The Second Law of Thermodynamics says that "in a closed system entropy never decreases" or, if you prefer, "in a closed system order never increases". By a "closed system", it meant an enclosed volume of space (such as the confines of a closed box) in which there is no exchange of energy and material with the outside. If, for example, the inside of the box has a non-uniform distribution of temperature (e.g. it is hotter at the bottom than at the top), then heat will flow until the temperature becomes uniform throughout the box. The entropy or disorder has thereby increased to its maximum. According to the Second Law, one will never observe the reverse (a decrease in entropy or increase in order) in which, for example, a "hot-spot" appears suddenly in a box with uniform temperature throughout. Perhaps the Second Law can be more easily visualized with a deck of ordered playing cards (e.g. reds on top and blacks on the bottom). Tossing the cards into the air and allowing them to fall randomly onto the floor will always disarrange the cards (i.e. increase the disorder or entropy). Since order never increases in a closed system, a shuffled deck of cards will never spontaneously rearrange itself such that reds are on top and blacks are at the bottom.

Many physical chemists point out the relentless increase in entropy. For example, Sir Arthur Eddington wrote: "The law that entropy always increases — the Second Law of Thermodynamics — holds, I think, the supreme position among the laws of Nature. If your theory is found to be against the Second Law of Thermodynamics, I can give you no hope; there is nothing for it but to collapse in deepest humiliation."

Now evolution describes a process in which simple life forms have, over millions of years, become increasingly complex. A mammal is far more complex ("ordered") than a bacterium. It is this fact that has led the extremists in the anti-evolution movement to conclude that evolution violates the Second Law of Thermodynamics. Henry Morris, for example, wrote: "The law of increasing entropy is an impenetrable barrier which no evolutionary mechanism yet suggested has ever been able to overcome. Evolution and entropy are opposing and mutually exclusive concepts. If the entropy principle is really a universal law, then evolution must be impossible."

To paraphrase the Morris argument: (a) Thermodynamics forbids decreases in entropy in a closed system. This is tantamount to saying the order cannot increase. (b) Evolution is a mechanism in which order is created because simple organisms are converted into complex organisms. (c) Therefore, evolution violates the Second Law, and evolution must be a false concept. What is wrong with this argument? Why does evolution not "collapse in deepest humiliation"?

The source of the confusion lies in the term "closed system" as expressed in my original statement of the Second Law: "In a closed system entropy never decreases." If, however, a system is not closed, that is to say if energy and material can move in and out of the system, all bets are off. Consider again the disordered deck of cards strewn on the floor. I can pick up the cards and, with the input of muscle energy from my arms, easily arrange the cards into all reds and all blacks. The result is a decrease in entropy because disorder has now been converted into order. But the Second Law has not been violated because, with the input of muscle energy, the system is no longer "closed". The Second Law of Thermodynamics is irrelevant.

Evolution likewise occurs in an open system: The Earth. The earth is an open system because it receives energy input from its ultimate source, the sun. Thus, the input of sun energy allows plant life to grow and evolve (with an accompanying decrease in entropy or increase in order). We and other animals eat the plants, thereby acquiring the sun energy indirectly. The plant energy is used by animals to grow and evolve (also with a decrease in entropy or increase in order). There may be problems and uncertainties with evolutionary theory, but violation of the Second Law of Thermodynamics is, most definitely, not one of them.

Part 6: The Gap Problem

Every once in a while it is beneficial to pause and reflect upon what has already transpired. Thus, the book began by describing in detail the tenets of natural selection theory. It then launched into various criticisms of natural selection, criticisms centered upon the evolutionary "story"; upon the validity of natural selection as a theory; upon wondrous features of Nature; and upon the concept of entropy. In each case, I have argued that the criticisms (while in some instances revealing the limitations of natural selection) do not, in the end, prove fatal to the theory.

I now address one of the most serious and widely invoked criticisms of natural selection: The *gap*. The criticism is founded upon the fact that many intermediate forms of life, expected from Darwin's gradual "descent with modification", are not found in the fossil record. I will not be coy by delaying my conclusions: I believe that although the existence of fossil gaps is self-evident, the gaps do not negate the relevance of natural selection to the evolution of life. This is because one can devise explanations for the gaps other than an inherent weakness in the natural selection mechanism. By the same token, the fossil record fails to provide the definitive and persuasive support for the imperceptibly small evolutionary changes that Darwin and his followers had originally postulated.

Darwin himself acknowledged that if species descended from other species by insensibly fine gradations, we should see innumerable transitional forms. He stated that his theory implied that "the number of intermediate and transitional links, between living and extinct species, must have been inconceivably great". In other words, if species D evolved from species A in a sequence A → B → C → D, then one might expect to find, for the evolution of D, fossil evidence for B and C (not just A). In order to drive home the concept of such "transitional links", I will begin with the best examples (or at least the examples most frequently cited by evolutionists) where intermediates have indeed been found. The three examples consist of the following sequences: fish-to-amphibian, reptile-to-mammal, and reptile-to-bird.

a) Fish-to-Amphibian

Land animals evolved from aquatic mammals. One would expect, therefore, the presence of intermediates between creatures that live in the water and creatures that live on land. Such animals do exist, and they are called amphibians. On an evolutionary scale, amphibians (such as frogs, toads, and salamanders) lie above fish but below reptiles (such as snakes and turtles). Amphibians lay their eggs in water where the young (called tadpoles) live much like fish, including breathing through external gills. Ultimately, the tadpoles undergo a metamorphosis during which they lose their gills and begin to breathe through lungs. In the course of metamorphosis, amphibians (with the exception of a few primitive species that remain limbless and worm-like) also develop legs which allow the adults to spend varying amounts of time on land. It seems certain that amphibians subsequently gave rise to reptiles, a group of animals that dispensed altogether with a gilled aquatic phase.

Several creatures living today are consistent with the idea that fish had transformed themselves into amphibians prior to the appearance of reptiles. For example, certain tropical fish, such as the Siamese fighting fish, possess lungs and must periodically swim to the water surface to breathe. Similarly, certain fish, notably the walking catfish, have lobe-fins that enable them to walk, albeit clumsily, across land. No far-fetched "story" is required to explain the adaptive value of this terrestrial fish locomotion: Walking catfish can thereby escape to a new pond should its old pond ever begin to dry out. Over time, adaptation such as lungs and legs were positively selected, and amphibians evolved. Despite the obvious lack of available details, it is not hard to imagine the following sequence: fish with gills → fish with primitive lungs → amphibians with gills in the young and lungs in the adults → reptiles with lungs only.

One can understand why evolutionists proudly point to walking fish, that can temporarily live out of water, as a likely type of intermediate in the evolution of amphibians. One can also understand the concern that, with gradualism as a cornerstone of natural selection, a greater number of fish-to-amphibians intermediates have not been found in the fossil record. The dilemma will be addressed below, but for the moment it is important to note that possible intermediates must not be visualized simply as "hybrids" of modern fish and modern amphibians. The scheme below clarifies the point. Assume that an ancestral fish (F1) evolved into a more highly developed fish (F3) bearing certain attributes, perhaps primitive lungs and legs, that led to formation, through a series of intermediates, to modern amphibians (A5). According to the model, fish F1 is also the ancestor to modern fish F7. Since an early fish precursor (e.g. F4) may appear quite differently from a modern fish, and since an early amphibian precursor (e.g. A2) may appear quite differently from a modern amphibian, one cannot easily "average" F7 and A5 (the only animals with which we are really familiar) in an attempt to guess what early intermediates F4 and A2 might have looked like.

	F1
	↓
Fish:	F2 → F4 → F5 → F6 → F7 (modern)
	↓
	F3
	↓
Amphibians:	A1 → A2 → A3 → A4 → A5 (modern)

The above scheme grossly understates the number of intermediates that, clearly, must have been involved in the transformation of fish to amphibians. There were not just a few intermediates but countless variations as the fish genome mutated over the eons to the amphibian genome. Herein lies the problem. We know that lungfish, for example, have fins and so-called "intestinal spiral valves" identical to that of any fish. The lungfish also has lungs and an "aerated-blood-return" similar to those found in terrestrial vertebrates. This is comforting to evolutionists since we have on hand a living specimen with traits that are neither 100% fish-like nor 100% amphibian-like. There exists, however, an absence of living animals with traits intermediate between fish and amphibian at the *organ* level (e.g. a fish with a partially developed lung). Opponents to natural selection demand hundreds, even thousands, of different species all unambiguously intermediate in terms of both organ anatomy and overall biology. These do not exist, plain and simple.

In summary, evolutionists point (correctly) to the lungfish as a likely intermediate organism of the type expected from an evolutionary mechanism. And opponents of natural selection point (correctly) to the rarity of organisms with partially formed transitional organs and transitional overall biology. Transitional gradations are definitely missing, a fact that did not escape Darwin who wrote, in a letter to his friend Asa Gray, that one's "imagination must fill up the very wide blanks." As will be seen momentarily, evolutionists have advanced possible reasons for the "very wide blanks", and a person's attitude toward natural selection depends in part upon his or her opinion of these explanations.

b) Reptile-to-Mammal

Monotremes, such as the duck-billed platypus, are another striking example of animals with intermediate features. The duck-billed platypus lays eggs like a reptile but has hair, warm blood, and mammary glands like a mammal. Monotremes may represent either a relic from the reptile-to-mammal pathway or, alternatively, an offshoot from a reptilian ancestor common to both monotremes and other mammal types.

Scientists generally agree that mammals did in fact evolve from reptiles. Let us assume, for the sake of simplicity, that reptiles needed to acquire only four traits to become mammals: hair, warm blood, live births, and mammary glands. (Actually, the list is much longer, but four traits suffice for illustration purposes.) It is possible to speculate, more by way of amusement than rigor, as to how reptiles transformed into mammals. As a start, one might postulate a sequential "one-trait-at-a-time" mode of development:

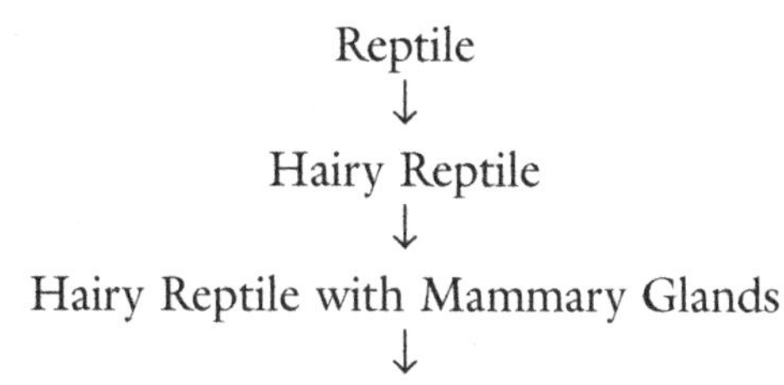

↓

Hairy Reptile with Mammary Glands and Warm Blood

↓

Hairy Mammal with Mammary Glands, Warm Blood, and Live Birth

A number of points can be made about this mechanism: (a) The exact order by which the four traits appeared during evolution is unknown, so that the sequence given above is totally arbitrary. (b) Likewise, the scheme involves intermediates, such as the "hairy reptile", for which evidence is non-existent (i.e. no "hairy reptiles" are alive today, and fossil records reveal little about hair because hair is so biodegradable). (c) The mechanism disguises the fact that there must have been a multitude of "secondary" intermediates (with partially evolved organs) interspersed between successive pairs of "primary" intermediates (with fully functional organs as given in the scheme). For example, the transformation of a "hairy" reptile into a "hairy reptile with mammary glands" might have taken place something like this (given here again in a highly condensed arbitrary format not to be taken literally):

Primary: Hairy Reptile

↓

Secondary: Hairy Reptile with Small Sweat Glands on Chest

↓

Secondary: Hairy Reptile with Large Sweat Glands on Chest

↓

Secondary: Hairy Reptile with Large Sweat Glands on Chest that Produced Liquid with Low Fat Content

↓

Primary: Hairy Reptile with Fully Developed Mammary Glands

Clearly, it required a substantial genome modification to create mammary glands on the chest of an animal that originally had no mammary glands. Such a genome would have been achieved only gradually because, according to neo-Darwinism, modifications were derived from one chance genetic modification at a time. Each such genetic modification constituted, of course, an intermediate of sorts. Not surprisingly, the scarcity of primary and (especially) secondary intermediates in Nature, among the countless numbers that must have existed once upon a time, provides fodder for opponents of natural selection. Since the gaps are unlikely to be ever filled, we have no choice but to live and deal with the problem, whether we like it or not.

At the other extreme of a "one-trait-at-a-time" mechanism, all four traits might have evolved more-or-less simultaneously:

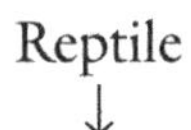

Reptile
↓
Hairy "Reptile" with Mammary Glands, Warm Blood, and Live Births (a mammal)

Such a process does not, despite its outward simplicity, preclude the need for a huge number of secondary intermediates. A hypothetical intermediate might, accordingly, have multiple changes all occurring in concert: Enlarged sweat glands on their way to becoming mammary glands; short bristles on the legs on their way to becoming hair over the entire animal; and blood that is not at a constant warm temperature (as in mammals) but warm only for short periods of time. Of course, I am free to allow my imagination to roam wildly because hard data on the subject are not available to contradict me!

Regardless of what particular prehistoric creatures are concocted, it is important to keep in mind the central principle of neo-Darwinism: Species evolution was propelled by that tiny percentage of random genetic modifications (including mutations) that, fortuitously, happened to be beneficial under the environmental conditions of the time. An "intermediate" represents a collection of such mutations, each mutation assisting in the formation of a partially developed organ or trait. But even if an organ were only partially developed, the "imperfect" organ must, nonetheless, have furnished survival value (or at least not been harmful) if the corresponding mutations were to be retained in the population. All genes directing the development of the various organs or traits within an intermediate must have mutated (beneficially of course!) again and again until, out of this incredible genetic melange, the reptile became a mammal.

Hair, incidentally, is thought to have first appeared on the scene as tiny sensor organs among the reptilian scales. When the sensors brushed against objects, the information was transmitted to the brain not unlike the action of cat whiskers or insect antenae. Perhaps the hair detectable on rat tails and armadillo shells, which have retained their ancestral scales, reflects a prehistoric intermediate.

Hair is an example of a structure that seemingly developed for one purpose (presumably, as mentioned, as a sense organ) and then later branched out into multiple other uses. Sea otters use air trapped in their fur to help them float. Capybaras have specialized hair under their tails that is covered with odoriferous scent; the hair is used to "paint" the forest floor with a sex-attractant trail. Water shrews have supportive foot hairs that allow them to run on the surface of water. African crested porcupines warn their predators by rattling their rigid tail hairs. The rhinoceros tightly compresses a wad of hair to construct its horn. And the sloth camouflages itself in the tropical foliage via poor grooming habits that allow green algae to grow on its brown hair. Most commonly, of course, hair is used for warmth in the form of fur. Thus, there is no reason why Nature cannot devise organs whose current functions are now totally different from those of past intermediates.

In all likelihood, the various traits in the reptile-to-mammal conversion evolved out-of-phase with each other. For example, hair and mammary glands might have originally developed side-by-side, but hair could have been "faster" and thus completed the job first. Considering how many traits had to be introduced during the evolution of mammals (all out-of-phase with each other), and how many individual small steps were required for each new trait, and how little information is available on all of this, one can easily become disheartened. There is, however, a noteworthy bright-spot in the story of mammal evolution, and this will be considered next.

Mammal-like reptiles, called theriodonts ("beast-toothed") or *therapsids*, constitute the "flag-ship" of fossil evidence supporting Darwinian ideas. The best known example, *Dimetrodon*, is commonly miscast as a dinosaur, but in fact it is actually a therapsid. *Dimetrodon* looked like an over-sized iguana except that extending upward, from its entire backbone, was a huge ribbed fan. Therapsids dominated terrestial life for millions of years before the rise of dinosaurs. Ultimately, therapsids lost out to dinosaurs and vanished, but not before a few of them managed to transform into the first mammals. Every cat, dog, cow, and human can, accordingly, look back to the therapsid as its ancestor.

Some therapsids were herbivores, some carnivores, while some omnivores. They ranged in size from that of a rat to a rhinoceros. T. S. Kemp wrote of the fateful transition from therapsid to mammal that occurred 215 million years ago: "This is one example known where the evolution of one class of vertebrates from another is well documented by the fossil record."

The term "mammal-like" in reference to therapsids does not signify possession of rudimentary hair, mammary glands, or warm blood (there is no way of really knowing whether such items were present or not...these attributes do not fossilize). It is the skull, particularly the jaw, that provides some of the most striking evidence that therapsids were mammals-in-the-making.

In reptiles, at the rear lower jaw near the attachment point to the skull, lies a group of small teeth-bearing bones called the "dentary bones". In mammals, there is only a single dentary bone connected to the skull, namely the lower jaw. Therapsid fossils found in South Africa, South America, and Britain show that the reptile-to-mammal transformation took place progressively in a sequence that corresponds, more or less, to the therapsid appearance in the geological record. The first thing to happen was that one dentary bone grew larger and crowded out the other smaller bones. Eventually, in the first mammals, two smaller bones of the reptilian lower jaw became incorporated into the middle ear (as the malleus and incus bones). These two bones, along with the stapes, gave mammals an acute sense of hearing over a wide range of frequencies. As A. W. Crompton said: "We hear with the bones that reptiles chew with. Why this took place, we have no idea. But the impact of all this is incredible."

A vestige of our reptilian heritage can be seen in mammalian embryos. In the early fetal stages, the *malleus* is part of the lower jaw. Only at a more advanced stage of embryonic development does the *malleus* move into the middle ear. Thus, our evolutionary history is "recapitulated" in the human fetus.

Therapsids were also evolving a mechanism for more complex chewing than the "up-and-down" motion characteristic of reptiles. The jaw was beginning to move forward-and-backward and side-to-side as well. In addition, therapsid skulls show a gradual increase in the size of the nasal passages, suggesting an enhanced sense of smell. But other than the altered skull, therapsids remained quite reptile-like. They had, for example, no enlarged brain, and their infant teeth were unsuitable for nursing. In summation, one has to be impressed with therapsids as evolutionary intermediates and as the definitive fossil evidence for the gradual evolution that Darwin originally postulated.

Most large therapsids probably died out in the so-called "Permian extinction". Many of the smaller remaining therapsids then fell victim to swift predacious dinosaurs that sprang up from another reptilian line. Fortunately for us, however, therapsids with mammalian traits (good hearing and smell), living as nocturnal creatures, survived in the shadows of the mighty dinosaurs. In the relative safety of their ecological niche, the creatures became more and more mammalian, destined to dominate the scene upon the demise of the dinosaurs.

So there they are, the remarkable therapsids, with skulls intermediate between those of reptile and mammal. Yet additional intermediates are curiously rare or absent, and therein lies the problem. Evolutionists see the intermediates; their opponents see mainly a void.

c) Reptile-to-Bird

Archaeopteryx, a reptile-like bird, is the best known example of a creature that seemingly lies intermediate between two great classes of animals. The first *Archaeopteryx* fossil, discovered by stonecutters in Bavaria only two years after the publication of the "*Origin*", came as a scientific bombshell. Darwin had already hypothesized that birds developed from reptiles, but he recognized that the absence of fossil intermediates constituted "probably the gravest and most obvious of the many objections which may be urged against my views." One can imagine, therefore, the delight of Darwin's supporters at the discovery of *Archaeopteryx*, a missing link that was to play an important role in documenting the evolutionist cause.

Five *Archaeopteryx* specimens are now available. The crow-sized animal had skeletal features of a reptile including a long, bony tail and wing-bones that terminated in three slender, unfused, clawed fingers. The wing-claws were presumably helpful in clambering about tree branches. *Archaeopteryx* was also non-avian in its lack of a sternum, a breastbone that anchors flight muscles. The head was lizard-like with the jaws having pointed reptilian teeth. But the arms, hands, and tail were adorned with modern-looking feathers, and this feature, more than any other, classifies the animal as a dinosaur-looking bird rather than a bird-like dinosaur. Only birds, after all, have feathers. In 1877, O.C. March declared that the gap between reptiles and birds had been bridged. *Archaeopteryx*, he said, served as the "stepping stone by which evolutionists of today lead the doubting brother across the shallow remnant of the gulf, once thought impassable." *Archaeopteryx* was, in brief, exactly the sort of intermediate form that Darwin's theory had predicted.

Scientists have still to reach a consensus on the ancestry of *Archaeopteryx* and, therefore, of birds in general. It is not even certain that the feathers of *Archaeopteryx* evolved for the purpose of flight. Alternatively, feathers may have originally developed to help some "proto-bird" conserve its body heat. *Archaeopteryx* has even been portrayed as basically a bipedal land animal that ran down insects for food; feathers were used primarily to beat insects out of the air like a fly-swatter. Although feathers are not particularly warm (compared to fur, wool, and down), and not endowed with holes to minimize air resistance as in an ideal fly swatter, Nature (as I have already amply stated) need not strive for perfection. A slight advantage will do.

Bird flight itself could have evolved either from "trees down" or from "ground up". Thus, it is easy to imagine survival value in wing-like structures that assist an arboreal animal in jumping from branch to branch and, ultimately, from tree to tree. The flying squirrel provides a good example. But, it has been argued, primitive wings might have, instead, allowed a terrestrial animal to take periodic leaps from the ground in order to catch food or to escape predators. There are also two schools of thought over whether *Archaeopteryx* was a feathered dinosaur or a bird. If it was a feathered dinosaur, as many paleontologists believe, then dinosaurs have managed to survive among us in the form of avian descendants like the robin, loon, and eagle (a delightful thought). Ornithologists on the other hand, claim *Archaeopteryx* as one of their own (and regard the dinosaur idea as "paleo-babble", to quote one of them). Despite the uncertainties, and despite all the unprovable theories regarding the evolution of powered flight, the existence of *Archaeopteryx* provides the most celebrated evidence for the one assertion on which almost all scientists agree: Birds evolved from some type of reptile.

Although *Archaeopteryx* may not constitute proof of natural selection, it is an excellent example of the sort of empirical observation that would remain mysterious in the absence of an explanatory construct such as provided by Darwin's theory of evolution. How have opponents of natural selection countered the palpable reality of this semi-reptilian bird (or bird-like dinosaur)? I list below some of the more common repartees:

a) It is not sufficient to find in the fossil record one or two organisms that seemingly lie intermediate between two groups. One would expect to find a whole series of transitional forms including, in the case of avian evolution, ancestral species with only partially developed feathers that could glide rather than fly. Intermediates with only partially developed wings would also be predicted.
b) Fossil evidence, in any case, deals only with skeletal features and necessarily ignores the 99% of biology concerned with the soft

anatomy. Birds and reptiles differ substantially in many physiological and anatomical characteristics (e.g. central nervous and cardiovascular systems), and no one has any notion of how *Archaeopteryx* fits into the evolutionary scheme with regard to such soft organs.

c) "*Convergent evolution*", a common phenomenon in biology, is defined as the development of identical traits among widely differing organisms. For example, vertebrates and cephalopods (e.g. the octapus) have remarkably similar eyes. The Tasmanian wolf (now extinct) is another widely cited example of convergent evolution. Although the Tasmanian wolf was a marsupial (like the kangaroo and opossum) and unrelated to the common dog, the Tasmanian wolf is almost indistinguishable from the dog family in gross physical appearance. By inference, *Archaeopteryx* may not be an intermediate at all but, instead, a bird that happened to acquire many of the same traits possessed by reptiles. Striking similarity, in other words, does not necessarily imply a close causal biological relationship.

d) Modern birds differ from reptiles in innumerable ways: The former have hollow bones and thin skulls; beaks instead of tooth-studded jaws; a lack of reptilian tail; an enlarged cerebellum and visual cortex in the brain; a crop in which to store food; a highly regulated body temperature; completed partitions in the heart; air-sacs within the body; a unique musculature, and so on. Since birds had 150 million years to first appear once life emerged from the sea, it is easy to imagine a particular trait evolving in such a long period of time. What is harder for some to comprehend, however, is the accumulation of the multitude of the avian traits integrated into a single functional pattern. As discussed earlier in the book, such incredulity may stem from an unfortunate dearth of hard information as well as from the basic limitations of human thinking when applied to extreme lengths of time.

I have discussed several classical examples of intermediate life forms. It is only fair to mention, however, that most evolutionary gaps are as unfilled today as they were in the time of Darwin. Some of the most important examples in this regard are the following:

a) The very fact that there exists a taxonomy (a classification system for plants and animals) proves that there is a discontinuity in the living world or, in other words, a presence of "gaps". Classification into species, families, phyla, etc. would be impossible if life were in fact blended into a smooth continuum mirroring its evolutionary roots. Gaps are a self-evident reality that must be explained in any evolutionary theory; I will attempt to do so at the end of this section.
b) In the Cambrian fossil record, one sees an explosive proliferation of life forms such as mollusks, jellyfish, sponges, and crustaceans. The creation of so many marine phyla in such a short time (only 15 million years) might, according to one theory, have been caused by a sudden increase in the oxygen content of the seawater. Strangely, however, virtually no ancestors of the Cambrian invertebrates are to be found in pre-Cambrian rock. Pre-Cambrian strata deposited for tens of millions of years prior to the Cambrian "explosion" are devoid of precursor fossils. Thus, a puzzling "gap" separates single-celled organisms (for which fossil evidence is plentiful) and the various complex Cambrian sea creatures (for which fossil evidence is also plentiful).
c) Darwin referred to the origin of flowering plants (angiosperms) as an "abominable mystery". (Darwin, it must be stated again, was unusually forthright, by modern standards, in admitting to the difficulties of his own theory. It is a source of great admiration). To this day, the mystery has not been solved. Flowering plants appeared, already highly specialized and diversified, in the Cretaceous period. No trace of their forebears have been found. It is even unclear whether flowering plants evolved from conifers or whether the two great categories of higher plants

arose from a common ancestor. The fossil cupboard is bare of transitional forms.

d) Evolutionists legitimately point to primate evolution (from tree shrew to lemur to monkey to ape to man) as an impressive sequential series. It has been claimed, however, that the "vector of progress" is far less impressive when examined via anatomical detail. For example, the conversion of quadrupedal (four-legged) primates into bipedal (two-legged) primates required substantial bony and neuromuscular changes in the pelvic girdle. Such profound changes could hardly have occurred by means of a single mutational event. Yet, with one possible exception just uncovered, no primate representing a transition between a quadruped and the first erect bipedal form is known. Darwin's model requires that, at some point, such an intermediate not only existed but that it possessed sufficient survival advantage to permit further progressive evolution.

e) Sharks and rays appeared abruptly 50 million years after most fish. If sharks and rays were preceded by a series of almost imperceptibly changing forerunners, as required by natural selection, paleontologists certainly can provide no hard fossil evidence for it.

It is hardly necessary to belabor further the obvious point that gaps are prevalent in evolutionary biology. The absence of fine gradations in the fossil record is the rule, not the exception. Thousands of different forms, all unambiguously intermediate in terms of their organ anatomy and overall biology, simply do not exist. Evolutionists (while understandably preferring to talk about *bona fide* cases of intermediacy) cannot — and in general do not — deny the "gap problem". Furthermore, few evolutionists predict that fossils yet to be discovered in the future will ever solve the problem. No one really has much hope, for example, that additional fossil hunting, no matter how intense, is likely to fill the pre-Cambrian void or to uncover a series of shark ancestors.

Consider now the equation below in which A = the original species; the I's = intermediates; and B = a new species that has evolved from A via the intermediates:

$$\begin{array}{l} 1\,000\,000 \text{ A's} \rightarrow 1\,000\,000 \text{ A's} + I_1 \\ \qquad\qquad\qquad\qquad\qquad\qquad\qquad \downarrow \\ \qquad\qquad\qquad\qquad\qquad\qquad\qquad I_2 \\ \qquad\qquad\qquad\qquad\qquad\qquad\qquad \downarrow \\ \qquad\qquad\qquad\qquad\qquad\qquad\qquad I_3 \\ \qquad\qquad\qquad\qquad\qquad\qquad\qquad \downarrow \\ \qquad\qquad\qquad\qquad\qquad\qquad\qquad I_4 \rightarrow 1\,000\,000 \text{ B's} \end{array}$$

The mechanism postulates that species A was originally in abundance (a fact denoted by a representative figure of 1 000 000). Suddenly, in some tiny locality tucked within the entire range of A, a modification I_1 appears. I_1 does reasonably well within its small domain, and over time it modifies itself further to I_2 and then to I_3 and I_4. Although these evolutionary intermediates survive among their more abundant relatives, their populations never really "take off" because there are not many of them to begin with. It takes time for a new creature to establish itself. Moreover, if an intermediate is not all that progressive (i.e. if their new traits, while not being outright disadvantageous, impart little in the way of survival advantage), then the numbers of this intermediate will be limited for this reason as well. Ultimately, it is assumed that the intermediates evolve into a new species B where the high reproductive advantage found in A is restored. (By "new species" one generally means an animal that can mate only within its own species). Perhaps B thrives and expands (note the arbitrary number "1 000 000" again) because it occupies a previously empty niche where competition is minimal. Or perhaps B thrives and expands because of some favorable new anatomical feature. Whatever the source of B's advantage is, the preceding mechanism is purely Darwinian in scope. The bottom line here is that species A has evolved into B via intermediates that never attain large populations.

A specific example will clarify the mechanism. Suppose, in the evolution of birds from reptiles, intermediates were formed in the following sequence: scales, hairy scales, stubby feathers, half-formed feathers, full-flight feathers. An intermediate with half-formed feathers might have only a slight advantage over bare scales. But one does not even have to postulate a slight advantage; the half-formed feathers may be simply "tolerated" if the energy costs for producing them are not debilitative. In any event, the intermediates, being mutational rarities and somewhat freak-like, never build up in population. They are simply there. This is not to say, however, that the intermediates are unimportant. Eventually, they lead to full-flight feathers which, no one will argue, had enormous impact. They allowed the animals, called birds, to take to the air and thrive.

Given the preceding scenario, what can paleontologists expect to observe millions of years after the fact? Since A and B were at one time widespread and present in large numbers, there is a small but finite chance that paleontologists will find a few of them that have become fossilized. The chance of a successful dig is small because numerous factors work against the creation and discovery of fossils: Most vertebrates did not die, presumably, from natural causes and then become (conveniently) buried under proper fossilizing conditions. Instead, they were eaten by bone-knawing and bone-crushing creatures. Only a small percentage would have escaped this fate and disappeared intact into the earth's crust. Given the huge expanse of the earth's crust, the chances are slim of ever finding a "burial site" and uncovering an interesting fossil. Recall that we now have only five fossils of *Archaeopteryx*! The only saving grace is that certain established species were plentiful in number, thereby enhancing somewhat the probability of discovering their fossils. When, however, it comes to those long sought-after intermediates, their limited populations and tiny geographical niches reduce the possibility of a fossil-find to practically nil. Hence the inevitable prevalence of "gaps".

So what is the conclusion from all this? In my opinion, the rarity of intermediates, especially at the organ level, is both puzzling and frustrating (the preceding rationale notwithstanding).

But puzzlement and frustration is an emotional reaction, not a rigorous scientific response. In purely scientific terms, one cannot exclude the possibility that the ubiquitous gaps have a plausible explanation within the context of Darwinian theory, and that, therefore, the theory must be accepted, or at least tolerated, until something better comes along.

Part 7: Complexity

The evolution of complex traits is one of natural selection's knottiest problems. Consider the vertebrate eye (a favorite organ of discussion in debates over the mechanism of evolution). Many who reject the fact that natural selection is the source of simple changes are, simultaneously, reluctant to accept the assertion that gradual accretion of random improvements among many separate but coordinated parts can generate anything as complex as the eye. To give a mechanical analogy, if one enlarges the size of a gear within a watch, every other part of the watch...other gears, spring, casing, etc.... must change their configuration accordingly for continued operation. Similarly, the multitude of eye-parts must have been modified over the ages in a coordinated fashion. "When I think of the eye, I shudder," said Darwin. He also wrote in "*The Origin*": "If it could be demonstrated that any complex organ existed which could not possibly have been formed by numerous, successive, slight modifications, my theory would absolutely break down". It is a tribute to the greatness of Darwin that he recognized, and honorably expressed, his puzzlement with the "organs of perfection". No alibis; no obfuscation; no hand-waving.

Although everyone has a notion as to the meaning of "complexity", the concept is in fact difficult to quantify or even define. Is a bacterium more complex than a computer or the Defense Department? Generally speaking, complexity measures the number of different parts all working within a coordinated, organized system. Perhaps in describing the relationship of natural selection to complexity, it is best to begin with an example of evolutionary change that is simple and understandable. I will then return to the eye problem.

Malaria, a disease that kills millions each year, is caused by a mosquito-borne *Plasmodium* parasite that invades the red blood cells. Certain people in the tropics have evolved an adaptation to counter the effect of the parasite: The sickle-cell trait. This trait is characterized by abnormal red blood cells that are crescent-shaped (rather than round) and spiny (rather than smooth). Parasites that attempt to attack the sickle-cell are impaled or otherwise damaged, so the person with the trait is thereby protected.

The sickle-cell trait is most beneficially inherited from only a single parent with the other parent not having the trait. If both parents possess the trait, their children can become "*homozygous*" for the altered gene. Children with such a double-dose of genes suffer from a serious anemia, obstruction of blood vessels, and other medical problems. So, as often happens in Nature, there is a trade-off. Part of the population (with only one gene) will be resistant to the deadly malaria, whereas another part (with two genes) will suffer from sickle-cell anemia. Overall, the trade-off must be beneficial since in certain sections of Africa up to 20% of the inhabitants have the sickle-cell trait.

In the 17th century, slave traders from Holland transported slaves from West Africa to the two Dutch colonies of Curacao (in the Caribbean) and Surinam (in South America). Since Curacao had no malaria, the sickle-cell trait conferred no advantages; indeed, the trait was outright harmful owing to its association with anemia. In Surinam, on the other hand, the sickle-cell trait afforded protection from the malaria that was endemic to the region.

Three centuries later (an extremely short period on the evolutionary time-scale), the descendants of the Curacao slaves show hardly any incidence of the sickle-cell gene, whereas the gene persists in the descendants of the Surinam slaves. This is a wonderful example of natural selection at work, and evolutionists are fond of citing it when expressing their support for natural selection.

No serious scientist would argue with this modern example of natural selection. It has been pointed out many times, however, that the sickle-cell trait represents only a single DNA mutation. As a result of this mutation, the red blood cells' hemoglobin (an oxygen-carrying protein) is altered slightly at a single site. (In chemical terminology, the sickle-cell syndrome is caused by a lone amino acid replacement within the huge hemoglobin molecule). As such, the formation and loss of the sickle-cell trait falls into the category of "microevolution", embodying the simplest of genetic changes possible. Unfortunately, microevolution teaches us little about the almost unfathomable sequence of "macroevolutionary" changes required to create a so-called "organ of perfection"—like the eye.

It hardly seems necessary to persuade anyone that the human eye is complex. The full extent of this complexity, however, can only be appreciated from information found in an anatomy text. Such texts list the various "eye-parts" as reproduced in the table below. Although most people, including myself, have little notion as to what these names mean, I can readily make the following claims: (a) The parts are sufficiently distinctive to warrant their own special names. (b) The parts, each with its own task, function cooperatively within an exceedingly complex system to produce a phenomenon called "vision". (c) Most people own the entire list of parts. They are inherent to the human genome, and they got there by evolution.

The table, which is by no means a complete listing, merits perhaps a ten-second scan. The reader will note the presence of a "levator palpebrae superioris". Do I know how a levator palpebrae superioris contributes to vision? No. Does anyone know how many genes control the formation of the levator palpebrae superioris? No. (In the fruit fly there are as many as 14 genes that modify eye color!) Does anyone know if there exists a single gene that impacts both the formation of the levator palpebrae superioris and other eye parts as well? No. Does anyone know the evolutionary history of a levator palpebrae superioris? No. Am I glad that I presumably have two functioning levator palpebrae superioris? Yes. This last affirmative response derives from my assumption that the absence or malfunction of most organelles listed in the table would cause serious eye problems. The eye, after all, is an organ-of-perfection.

Actually, the list of eye-parts in the table grossly underestimates the eye's true complexity because it ignores structures at the molecular level. Vision involves a series of biochemical reactions and neurological events that contribute even further to the intricacy of the visual process. A complete list of eye-parts, therefore, would include items such as retinal, rhodopsin, transducin, arrestin, and all the other biochemical constituents of the eye.

In the absence of hard information on the human eye (and on the eyes of sighted organisms from which we have descended), any discussion of eye evolution is necessarily superficial. It is surprising how heated the debate over eye evolution has become when both sides of the issue know so little!

Table. Parts of the Human Eye and Supporting Structures.

Superior rectus	Choroid
Orbital fatty tissue	Optic disc
Optic nerve	Dura mater
Inferior rectus	Canal of Schlemm
Levator palpebrae superioris	Ciliary muscle
Superior conjunctival fornix	Optic tract
Superior tarsus	Lateral geniculate body
Inferior tarsus	Optic radiation
Inferior fornix	Visual cortex
Inferior oblique	Ciliary nerves
Superior oblique muscle	Oculomotor nerve
Medial rectus	Ciliary ganglion
Trochlea	Accessory oculomotor nucleus
Tendon of superior oblique	Lacrimal gland
Cornea	Vorticose vein
Optic chiasma	Lacrimal nerve
Anular tendon	Lacrimal artery
Levator palpebrae superioris	Medial palpebral ligament
Lateral rectus	Tendon of superior oblique
Zygomatic bone	Orbital septum
Ethmoidal arteries	Inferior tarsal muscle
Posterior ciliary arteries	Inferior lacrimal canaliculus

(*Continued*)

Table. (*Continued*)

Ophthalmic artery	Lacrimal sac
Central retinal artery	Supraorbital foramina
Retinal arteries	Semilunar fold of conjuntiva
Supratrochlear artery	Lacrimal punctum
Supraorbital artery	Nasolacrimal duct
Anterior ciliary artery	Infraorbital nerve
Iridial arteries	Angular artery
Majory arterial circle of iris	Aponeurosis
Iridial fold	Maxillary sinus
Minor iridial ring	Iris
Major iridial ring	Lens
Sclera	Conjuctiva
Superior temporal artery	Ciliary Body
Optic disc	Zonular fibers
Inferior macular artery	Ora serrata
Inferior temporal artery	Vitreous body
Fovea centralis	Retina

An eminent Darwinist has ridiculed his opponents' tendency to "argue from personal incredulity" by his publishing a list of phrases commonly found in the anti-Darwin literature: "I find it hard to understand..."; "It does not seem feasible to explain..."; "I cannot see how..."; "neo-Darwinism seems inadequate to explain..."; "How could an organ so complex evolve..." etc. As I have myself stated earlier, personal amazement should never be a major component of a scientific argument. Nonetheless, I forgive those (including Darwin) who violate this rule with regard to the human eye. If no comprehensive theory is available to explain the eye, then someone contemplating the eye can, understandably, express astonishment at the eye's complexity. Such emotional responses are acceptable in scientific writing when, for all practical purposes, they are synonymous with a perfectly valid scientific statement such as "Current theories do not explain..."

I, personally, am perplexed at the evolution of the eye. Most scientists who have not entirely lost their childish sense of wonder probably feel likewise. But, out of respect for those who dislike emoting, I will recast my thoughts in more scientific terms: The evolution of the eye cannot be "explained" by neo-Darwinism because the theory cannot yet handle extreme complexity coupled to ignorance of genetics and developmental biology. This does not mean that neo-Darwinism is incorrect. It does mean that there are limitations as to how far the theory can be profitably extended given our present level of understanding.

Hampered by insufficient information, confounded by conflicting arguments, annoyed by diatribe, and disheartened over the prospects of ever understanding the evolution of complex organs, one feels (as did Darwin) like shuddering. I would drop further discussion of this eye business altogether were it not for the instructive, and even amusing, manner in which others have confronted the problem. I reiterate that the problem here is not about the *fact* of eye evolution but, rather, the *mechanism* of eye evolution.

Nobelist Peter Medawar once wrote: "It is silly to be thunderstruck by evolution of organ A if we should have been just as thunderstruck by a turn of events that had led to the evolution of B or C instead." Let us now examine the merits of Medawar's intriguing assertion by using the game of poker as an analogy.

The probability of any given poker hand is 1 in 2 598 960. Now suppose a player receives the following hand: Four aces and a king of hearts. There are two possible responses to this hand: "I am thunderstruck", or "It is silly to be thunderstruck because, while the probability of my wonderful hand is only 1 in 2 598 960, this is the identical probability for any other particular hand one might name." Which is the more sensible emotion, delight or boredom? It is certainly true that "four aces and a king of hearts" is equally probable to any specified useless hand (e.g. a two of hearts, a four of hearts, a six of spades, an eight of clubs, and a nine of diamonds). Nonetheless, a "four aces and a king of hearts" is an amazing hand because, according to poker rules, such a hand has been deemed desirable. If the above "two, four, six, eight, and nine of mixed suit" had also been designated as desirable, then this hand would be no less amazing than the "four aces and a king". One can be unabashedly amazed by "four aces and a king of hearts" because its probability, relative to the probability of all useless hands taken collectively, is exceedingly small.

How does this relate to Medawar's claim that any "awe of the eye" is unwarranted? Just as we have defined "four aces and a king of hearts" to be a desirable arrangement of cards, we can regard "the eye" as a desirable arrangement of innumerable eye-parts. The latter arrangement is desirable in the sense that it produces vision. One need not be amazed by the eye if "any odd arrangement" of the eye-parts will likewise generate vision. But if only a few arrangements of eye-parts (out of an indescribably large number of useless combinations) results in vision, then the achievement of such organization is indeed worthy of our admiration. I favor the latter possibility because, of one thing, I am persuaded that rearranging my eye-parts, even slightly, would indeed affect my sight adversely.

Glaucoma, which afflicts some two million Americans, is the leading cause of blindness. The condition is caused by improper drainage of the aqueous humor, a fluid that is being continually produced in the front part of the eye. Instead of passing through the "trabecular meshwork" and returning to the bloodstream, the fluid accumulates and creates an intraoccular pressure. Since the pressure slowly destroys the retina, blindness can result. The disease is controlled by, among other methods, topical administration of a drug called a "beta-blocker" that impedes the output of fluid. The cause of the drainage problem is not fully understood because there is no overt anatomical feature that can be seen to block the fluid. Blindness is caused by more subtle biochemical factors, and it is treated on this basis as well. Glaucoma is a good example of how every eye-part (whether a visible entity as listed in my table or one of the innumerable biochemical components not listed in the table) must work in perfect harmony.

Evolutionists do their cause no good when they imply, as did Medawar, that organs of extreme perfection are a humdrum evolutionary product hardly more noteworthy than the color of a moth's wing. Few will be fooled by this. The eye is a marvelously complex and highly improbable combination of vastly different parts.

Natural selection, proceeding over an almost unfathomable time-span, has no doubt been a contributing force in the formation of such an improbable structure. But this is sparse and unsatisfying information. What are the details with regard to the human eye evolution? Again, it is best that evolutionists are forthright and unhedgedly truthful. They can point to more primitive sight organs in simpler organisms but, with regard to the actual step-by-step evolution of the human eye, they can only say: "We don't know. We may never know. Nature may have neither given us the intellect nor the information necessary to understand all aspects of biological complexity." These are, of course, difficult words for many scientists to express.

With so little hard data in hand, scientists are free to speculate without fear that a convincing and embarrassing counter-argument will appear on the scene to spoil our day. Strict Darwinists feel that the eye did indeed evolve in a series of almost imperceptibly small steps. It is claimed that 1% of an eye is better than blindness; 5% of an eye better than 1%; 10% better than 5%, and so on. With each improvement in "percentage eye", the organism had an enhanced likelihood of surviving and reproducing. Over millions of years, organs of sight became more and more sophisticated and specialized until the eyes of higher animals were in place. In support of this model, one can point to the ascending order of complexity within the animal kingdom, starting from the simple pigmented cells ("eye spots") of primitive organisms all the way to the elaborate eyes of vertebrates. Although the sight organs of lower animals are not necessarily ancestral to vertebrate eyes, the former do indicate that simpler designs are indeed capable of functioning at some level useful to the organism.

It is instructive at this point to insert an argument for Darwinian evolution for the human eye that seems more philosophical than biological. Let us assume that the human eye could not have evolved in a single step from no eye at all. This means that the eye must have evolved from some related organelle that will be called "precursor-1". I now claim that a single mutation created the modern human eye from its most recent forerunner: Presursor-1.

A skeptic retorts, however, that precursor-1 is too different in structure from the modern eye for this to have happened. Fine. I then change the structure of precursor-1 to make it resemble the modern eye even more. If the skeptic balks again, I continue to make precursor-1 progressively more similar to the modern eye until a point is reached at which the skeptic finally concedes. Of course, where this point lies depends upon the intransigence of the particular skeptic. But once victory over the skeptic is achieved, I can then propose a precursor-2 that preceded precursor-1 and repeat the process. Ultimately, the skeptic will agree to a continuous series of precursors that connect the human eye to an unlikely looking ancestral organ or, perhaps, to no eye at all. If the differences between the successive precursors are made sufficiently small, the mutations necessary to affect the changes would almost certainly be within the realm of possibility.

So much for the strict Darwinists. What arguments have been levied in opposition? It has been pointed out, first of all, that there exists no evidence for a series of structures expanding slowly and smoothly in complexity until, finally, the human eye is reached. Thus, the idea of hundreds, or perhaps thousands, of precursors leading up to the human eye is mere speculation. Flaunting the "eye-spot-to-vertebrate-eye" sequence is misleading because few of these different types of eyes in the animal kingdom are thought to have evolved from each other. Each seemingly evolved separately via its own set of (unknown) intermediates. Even if a moderately comprehensive sequence of intermediates were available, we would still not necessarily understand the mechanism of change. For example, the fact that the forerunners of fish, with their single light-sensitive organ, evolved into fish with paired eyes does not tell us what took place mutationally and embryologically to create two eyes from one.

To continue the counter-argument against Darwinistic claims: It was proposed above that "5% of an eye" is better than "1% of an eye". But is it not just as likely that 1% and 5% eyes are equivalent, i.e. both useless? One could imagine even a "95% of an eye" that is totally non-functional (e.g. an eye in which all eye-parts are present and working except for a defective retina). If the human eye evolved from 1000 precursors, then, according to Darwinian doctrine, precursor-795, for example, must have provided better vision than precursor-794 which, in turn, must have provided better vision that precursor-793, and so on down the line. If some precursor is ever less favorable to survival than the one preceding it, the evolutionary line of descent is broken. By postulating an endless stream of precursors (to make structural changes as small and palatable to skeptics as possible), Darwinists must also accept that the survival benefit of each change is incredibly small. One then has to be concerned about the time required for such a tiny beneficial trait to take hold in the population before the corresponding gene is spontaneously eliminated by random mutations. And large or small, the structural changes would have to occur in an unfaltering, coordinated improvement among a litany of exquisitely arranged eye-parts. How was this accomplished when, similar to a slight increase in the size of a watch's gear, one part affects the suitability of all its neighboring parts? Natural selection theory can "explain" the eye only by becoming as complex and incomprehensible as the eye itself.

I have attempted to argue here both sides of the issue as concisely and persuasively as possible, no doubt creating a degree of confusion among the readers. Such confusion is normal and justified because, to repeat a previous point, we lack the information to formulate any definitive explanation for organs of extreme perfection. My own position is one of ambivalence. I share the frustration of those who wish for what we will probably never possess: Hard evidence for intermediates that are preserved and amplified because each one represents, in an increasing order of effectiveness, a beneficial function. On the other hand, I find many discourses attacking natural selection to be nihilistic. It is far easier to reveal to the world the limitations of a theory than it is to replace it with something better. I happen to believe that natural selection, a proven mechanism for microevolutionary changes, also played a role in the development of the human eye. Whether natural selection was the only mechanism, or even the major mechanism, I do not know. Given this uncertainty, I feel that it is just as misleading for Darwinists to tell the public that we understand more than we do as it is for the opponents to tell the public that we understand less than we do.

Thus far the human eye has been extolled as the paragon of biological complexity, but it should be added that even seemingly trivial evolutionary modifications, such as the famous lengthening of a giraffe's neck, demand a host of interrelated anatomical changes. In order to attain a longer neck, the giraffe had to evolve the genetic wherewithal to: Extend the neck vertebrae and all the associated muscles and tendons; decrease the weight of the head; modify the circulatory system to pump blood to greater heights and to reduce the blood pressure "rush" to the head when the animal drank; expand the lung capacity so as to expel the longer column of stale sir; change the skeletal framework including the lengthening of the forelegs; render the skin more impermeable to fluid loss from hydrostatic pressure; and create new postural reflexes to help escape from predators. Evidently, many synchronized mutations were required to create a longer neck, the relative "simplicity" of the trait notwithstanding.

A subtle but important point about Darwinian evolution relates to what has been termed "*iterative randomization*". Consider the following Darwinian mechanism for converting 5 animals (all with trait A) into 5 animals (4 of which have trait C):

	1 Mutation	**2 Selection**	**3 Mutation**	**4 Selection**	**5**
Animal no. 1	A	A	B	B	B
Animal no. 2	A	B	B	B	C
Animal no. 3	A	A	A	A	C
Animal no. 4	A	A	B	C	C
Animal no. 5	A	A	B	B	C

In the beginning (state 1), all 5 animals have trait A. A beneficial mutation then occurs in Animal no. 2 to transform an A into B to give state 2. Since trait B is advantageous compared to trait A, natural selection serves to enhance the relative population of trait B (state 3). Next, a second beneficial mutation in Animal no. 4 converts a B into C to give state 4, and herein lies the key point: The probability of obtaining a C-trait is enhanced by virtue of natural selection having previously increased the frequency of trait B from which trait C derives. If there had been no natural selection (i.e. if state 2 had not become state 3), then a "B-to-C" event would have been far less likely since there is only one B in state 2. Evolution may have no foresight, but it is clearly guided by the past. Those traits that are the most profitable will spawn the greatest number of progeny and, consequently, promote the likelihood of continued evolution. The only problem with this "iterative randomization" concept is that it implies that intermediates (B in the example) should be plentiful, but, as we have already seen in the discussion of gaps, intermediates are seldom found in large numbers.

Complexity is not confined to the organ level but extends even to multi-organism systems, as is well illustrated by the brainworm (*Dicrocoelium dendriticum*). This worm requires sheep, snails, and ants in its life cycle. The worm lives and reproduces in sheep gut. The sheep feces, containing the worms, are eaten by snails that, in turn, expel larvae in the mucus they produce. The mucus is a favorite food of the third brainworm host, a species of ant. Since ants are seldom eaten by sheep, entry of the worm into the ant would do the worm no good in completing its cycle were it not for the fact that an occasional worm finds it way into an ant's brain. In a totally uncharacteristic act, the deranged ant then climbs onto a stem of grass where it waits until it is ultimately eaten, along with the grass, by a sheep. And the cycle begins again. No one has ever hazarded a guess as to how this coordinated multispecies cycle ever came into existence.

Although many biologists believe in an evolutionary drive toward ever increasing complexity, complexity in and of itself does not necessarily confer an advantage. In fact, in certain instances evolution seems to have moved in the direction of diminished complexity. For example, a comparison of modern and fossil mammalian backbones suggests that changes toward greater and lesser complexity have occurred at about equal frequencies. Loss of eyes in cave-dwelling fish might also be viewed as a macroevolutionary simplification. Thus, an evolutionary push toward increased complexity should not be regarded as an ironclad rule.

Complexity may have its origins, to an unknown extent, in so-called "kin selection". Kin selection is based on the observation that many animals sacrifice their lives in order to save their kin. Examples range from the suicidal attack on an intruder by honey bees, to a man perishing while rescuing his children from a burning house. One might suspect that such altruistic behavior, and the corresponding genes tending to promote it (if such exist), would get eliminated if the possessor killed itself and thus prospects for passing on the trait. But kin selection theory points out that the savior and the saved are often relatives with common genes. The family genome, including the tendency toward altruism, can be preserved and even expanded despite one particular member losing his life in a successful act of heroism.

Kin selection can be applied to a crucial step in the evolution of complexity: The transition from single-celled organisms into multicellular animals or plants. It is reasoned that neighboring single-celled organisms tend to be related, i.e. have genes in common. If one such cell assisted the survival of its neighbor, even at some cost, then this would benefit the entire intercellular subset of genes. One could imagine, for example, a cell exuding a tiny amount of predator repellant — an amount too small to do much good. But a group of single-celled organisms living together within a confined volume of space might collectively exude sufficient repellant to constitute a definite survival advantage. Ultimately, it made sense to cooperate to a point of forming a unified community or, in other words, a multicellular organism. Given enough time, kin selectionists would argue, complex and even intelligent life was almost "inevitable". Of course, the leap from cell aggregation to intelligent life lacks the detail, the information content, and the breadth of explanatory power to truly satisfy the inquiring mind. Nonetheless, cooperation and self-sacrifice for the common good among kindred cells and kindred animals must, without question, be included as a component of evolution just as gene exchange among symbiotically related organisms must be included.

I should not end this section on complexity without mentioning the "*macromutation theory*" as advanced by Richard B. Goldschmidt, who was a first-rate intellect, over five decades ago. Macromutationists had trouble accepting the neo-Darwinian premise of organ creation via the accumulation of innumerable micromutations. Most notably, they could not understand why such organs, in their initial stages of evolution, would be preserved in the population before the primitive structures could (ostensibly) exert any positive benefit. It was proposed, as an alternative, that every once in a great while there appeared a "macromutation" associated with a rather massive structural modification. Although the vast majority of such changes would be "freakish" and thus disappear, a few such macromutations would have a dramatic and disproportionate impact on the overall course of evolution. Thus, macromutationists are believers in natural selection; their disagreement with neo-Darwinists (including punctuationists) stems largely from different views on the abruptness of structural changes.

Goldschmidt's 1940 book entitled "*The Material Basis of Evolution*" elicited a great deal of negative response. He wrote: "The neo-Darwinists reacted savagely. This time I was not only crazy but almost criminal." Later he wrote: "I am confident that in twenty years' time, my book, which is now ignored, will be given an honorable place in the history of evolutionary thought." This has not happened, not yet anyway, and it is instructive to explore briefly why not.

Modern geneticists regard macromutational transformations (e.g. a sudden conversion of a reptilian scale into an avian feather) as totally unacceptable. Such gross structural changes no doubt require the mutation of several genes, and it is unlikely, indeed impossible, that all these mutations could take place simultaneously. Beneficial mutations in a single gene, let alone in each member of a gene family, are rare occurrences.

Macromutationists, those few who are still holding the fort, have responded to the familiar criticisms by noting, first of all, that their mechanism explains the existence of gaps in the fossil record. Few intermediates are ever found because their formation is not an essential element of the macromutational model. The probability issue is handled in embryological terms. Many mutations that produce only minor alterations in the early embryo are known to cause extensive changes in the adult. For example, a mutation may alter a glycoprotein on the surface of an early embryonic cell and, thereby, affect the cell's propensity to migrate; this could have massive implications for the adult later on. Thus, development and differentiation in an embryo seem to amplify the effect of certain mutations. Macromutationists, in summary, propose that evolution is proceeded by this type of amplified mutation as opposed to the accumulation of many slight variations.

Both the neo-Darwinists and the macromutationists invoke natural selection, and it is quite possible that both mechanisms contributed to the evolutionary process. Never underestimate the antagonism possible between the two groups whose philosophies are similar but do not overlap precisely! One thing seems clear. Our ignorance in embryology is a serious impediment to a broader understanding of evolutionary biology. Progress in evolutionary thought awaits progress in developmental biology, pure and simple.

In concluding this section on complexity, I must admit that the arguments have gone back and forth, and none of them seems totally satisfying. Complex organs (as with the gaps in the fossil record) are a puzzle, but one can hardly dismiss natural selection on this basis. Consider, for example, a theory that has been developed to explain two observations, A and B. If both observations are explained, the theory must obviously be considered viable. If either observation A or B contradicts the theory, the theory must be discarded. Finally, if observation A is explained, while observation B lies "in a state of limbo" owing to lack of information or inadequate power of the theory, then the theory must still be retained. One might claim that the theory is weak or limited in its scope, but the theory has not been directly falsified. Natural selection falls into this last category. It explains beautifully microevolutionary events such as antibiotic resistance in bacteria and sickle cell anemia. It cannot fully deal with the human eye because the complexity of the eye supersedes the capabilities of the theory; because information on intermediates is missing; and because the ability of the human mind to comprehend events occurring over a vast time span has its limitations. But since the presence of the human eye does not contradict outright natural selection as an evolutionary mechanism (a mechanism in which the survival advantage of good eyesight is obvious), the theory has not been invalidated. Natural selection, despite its problems, must be considered valid until proven otherwise. In the second half of the book, I shall demonstrate an important weakness in natural selection as an all-encompassing theory.

Part 8: Molecular Evolution

The advent of molecular biology in the 1950s brought fresh hope and excitement to the field of evolution. Evolutionary pathways could now be traced not through gross anatomical features but through precise structures of DNA and proteins. No survey of Darwinism can be considered complete without at least a brief discussion of how DNA and proteins have become modified in the course of evolution. Those with limited interest in chemistry have no fear — there is no plan to introduce here complicated chemical structures. Molecular evolution can be explained and understood with only the barest of chemical detail. Nonetheless, readers with an aversion to anything "molecular" could well skip this short section without sacrificing an appreciation of evolution as provided in previous pages.

Genetic material (i.e. the DNA in cell nuclei) consists of extremely long chains in which four units, and only four units, are hooked together in a linear sequence. The four units are designated as A, T, G, and C. Thus, a tiny section within a DNA chain might appear as ...A-A-G-C-A-T... or ...A-T-C-G-C-T... where the dotted lines indicate that the chains continue, with an unspecified sequence of units, both to the right and to the left of the segments.

Actually, DNA is double-strand (the famous "double helix") in which two chains are held together by weak attractive forces. An important rule determines the construction of a DNA double-strand: A and T are weakly attracted to each other, as are G and C. Thus, if we know the sequence of one chain, we know the sequence of its partner chain:

```
DNA chain 1      ...A-A-T-C-G-G-T...
                    :  : : :  : : :
                    :  : : :  : : :
DNA chain 2      ...T-T-A-G-C-C-A...
```

The vertical four dots represent the weak attractions between A & T and G & C in the double-strands comprising the chromosomes in our cells' nuclei.

In cell division, the weak attractions are broken, the two DNA chains separate from each other, and the chains wander to opposite ends of the cell. Each chain then serves as a "template" or pattern with which to rebuild a partner chain. For example, chain 1 in the above sequence would attract the correct individual units (floating around as single molecules within the cell) as dictated by the A & T and G & C rule:

```
                                      ...A-A-T-C-G-G-T...
                                         : : : : : : :
...A-A-T-C-G-G-T... + A, T, G, and C →   : : : : : : :
                                      ...T T A G C C A...
```

Afterwards, the units are joined to each other (via formation of new chemical bonds represented by a dash between the units), and reconstruction of the original double-strand is now complete:

```
...A-A-T-C-G-G-T...
   : : : : : : :
   : : : : : : :
...T-T-A-G-C-C-A...
```

The same process occurs at the other end of the cell. It remains only for the cell to pinch into two at the middle to form two distinct cells that are genetically identical to the original parent cell. Cell division is now complete.

Having "disposed" of DNA chemistry, it is now possible to deal with proteins. Proteins like DNA, are long chains, but this time there are about 20 building blocks (called amino acids) that are linked together linearly. I will refer to the 20 units as P_1, P_2, P_8, P_{17}, etc. Thus, a typical protein might have, say, 300 amino acids, a small section of which is shown below:

$$...P_{19}\text{-}P_2\text{-}P_5\text{-}P_5\text{-}P_{17}\text{-}P_2\text{-}P_8...$$

What is the function of proteins? The most abundant protein, muscle, is well known to everyone as "meat". Other proteins comprise hair, tendon, and nail. But proteins also perform a less obvious function critical to life. Certain types of proteins, called "enzymes", catalyze the various biological reactions occurring in the cells. Almost every reaction in biology has a specific enzyme associated with it that allows the reaction to occur at very high speed. Were it not for enzymes, biological reactions that are completed in seconds could take years to accomplish. In effect, we are what our enzymes allow us to be.

Now comes the most important principle in modern biology: The sequence of units in the DNA determines the sequence of units in proteins. In other words, genetic material (i.e. sequences of A, T, G, and C) exists in large measure to dictate the structure of our proteins (i.e. sequences of amino acids) including enzymes. The principle is expressed in a wonderfully simple equation:

$$\text{sequence of DNA units} \rightarrow \text{sequence of protein units}$$

A "gene" is simply a long section of DNA that controls the sequence of one particular protein. Although much is known about how DNA sequences control protein sequences, the details here are unimportant. Suffice it to say that, ultimately, three DNA units "code" for one protein unit. For example, an A-C-A triplet within the DNA may for code for P_5, whereas an A-A-A triplet may code for P_8. After the P's are lined up, in a sequence determined by the DNA "template", the P's are joined together to form a specific protein. The important point to remember is that: One gene gives one protein.

It follows from the above that an error in DNA (caused perhaps by an X-ray or a toxic chemical) can lead to a modified protein. For example, if the A-C-A triplet is mutated into an A-A-A triplet, then P_8 will be incorporated into the protein instead of P_5. The consequences to a cell of such an enzyme modification range from nil to fatal. A 300-unit enzyme possesses certain amino acids that are intimately involved in the catalytic process. Loss or replacement of one of these would almost certainly have dire consequences to the cell because the cell would then lack an important biochemical reaction. On the other hand, many amino acid units are only distantly related to the actual site of enzyme-catalyzed reaction, and their loss or replacement may not affect the enzyme's catalytic ability.

Two final points must be made before DNA/protein chemistry can be applied to evolution. (a) A DNA mutation that damages a somatic cell (a non-reproductive cell as those in skin and bone) usually does not harm a multicellular organism (unless the damaged cell turns into cancer). If, however, a harmful mutation occurs in a "germ cell" (a reproductive cell, e.g. ovum or sperm) then all the cells in the progeny resulting from that germ cell will be adversely affected. Serious or fatal genetic disorders will result. (b) DNA modification of germ cells drives evolution. When an X-ray or chemical or virus or symbiotic organism modifies the DNA, a variant appears. Most of these externally induced alterations of DNA (and the corresponding protein) are disadvantageous to the organism. One cannot randomly change the structure of a gene and expect improvement (just as one cannot randomly tinker with a clock and expect improvement). Only rarely will germ cell modifications result in a more viable organism, but it is this rare event upon which evolution depends. The evolutionary mechanism is extraordinarily inefficient, and it functions only because of the vast time periods available to it.

Molecular biology has added a new dimension to the study of evolution. By comparing DNA and protein sequences of two species, one could hope to quantitate the "genetic distance" between them. Until the development of molecular biology, evolutionists had no clear and consistent yardstick by which they could assess the extent of an evolutionary change. Relating a series of species (e.g. deer, antelope, and elk) to points on an evolutionary clock depended upon subtle anatomical differences that a non-expert often could not appreciate. Even among experts, disagreements as to what was related to what were common. Molecular biology, on the other hand, had the ability (or so it seemed years ago) to change all that. Who could argue with sequence data from a blood protein showing that species A was much more similar to species B than to species C?

Let us begin with a particularly interesting comparison between humans and their nearest relative, the chimpanzee. Recently, the complete gene sequences of both species have been deciphered. Thus, amino acid sequences of hundreds of human and chimpanzee proteins are known. One can now readily examine, amino acid-by-amino acid, the differences between corresponding proteins from the two species. Recall that proteins are, typically, 100–800 amino acid units long, each unit being one of 20 possible amino acid structures.

Although it is not convenient to compile here more than a fraction of the available information, a few representative data will serve the purpose. Many key proteins with fancy names (e.g. fibrinopeptides A and B; cytochrome C; lysozyme; hemoglobin α and β) are identical in humans and chimpanzees. Various other proteins manifest small differences (myoglobin, 1 out of 153 units; carbonic anhydrase, 3 out of 264 units; transferrin, 8 out of 648 units). All in all, the sequences of human and chimpanzee proteins examined to date are, on the average, about 99% identical. Differences in proteins between humans and chimpanzees are seemingly too small to account for the obvious differences we see between the two species.

If the obvious biological uniqueness of man and chimpanzees is not based on protein sequences (and their corresponding DNA sequences), then from where do their characteristics derive? The most likely explanation is that humans and chimpanzees possess major differences in their so-called "*regulatory genes*" — those genes that control whether other genes are expressed or lie dormant. This is an unpleasant conclusion because much less is known about the sequences of regulatory genes and their mode of action. Quantitation, as can be accomplished by protein sequencing, is less meaningful with regard to regulatory genes. For information on evolutionary pathways, we might well return to the taxonomists who, in many cases at least, are able to detect more significant differences among species than can the biochemist!

In summary, species diversity is not directly attributable to changes within structural proteins. Anatomical differences seem to have arisen mainly from mutations affecting gene expression especially during embryonic development. Until such time that one understands the principle governing gene expression, and how it varies from species to species, molecular genetics will have only limited success in solving evolutionary problems.

The similarity, and in some cases identity, of human and chimpanzee proteins reflects the unity underlying the diversity of life. All life utilizes, for example, DNA, protein, sugars, and lipids. Many complex organic molecules are distributed widely among diverse species (e.g. chlorophyll in all plants; hemoglobin and testosterone in most vertebrates). Such close chemical kinship in Nature demonstrates that species and molecular evolution have been, to a substantial extent, independent from each other. It is, therefore, difficult to deduce information about one process from the other. The hopes and expectations originally accorded to molecular evolution have, consequently, not yet been fully realized.

Despite the shadow cast upon molecular evolution, it is well worth to examine the ingenious attempts to trace evolutionary lineages via DNA and protein structure. Take, for example, ribonuclease (an enzyme secreted by the pancreas) whose sequence is known for three dozen species. An evolutionary tree was constructed by using the method of "maximum parsimony". In effect, the protein sequences were, with the aid of a computer, arranged such that those with the fewest discrepancies were near each other on the tree. This was not a trivial exercise because one had to be alert for deletions, restored mutations, and other complications as illustrated by segments from two corresponding proteins produced by two different species:

Species 1: $...P_2$-P_9-P_3-P_7-P_6-$P_8...$
Species 2: $...P_2$-P_3-P_7-P_6-P_8-$P_7...$

This would, on the surface, appear to be serious mismatch (five out of six) between the species until one realizes that Species 1 had probably experienced a deletion of P_9 between P_2 and P_3. Matching is excellent if this deletion is taken into account and Species 2 is resupplied with a P_9:

Species 1: $...P_2$-P_9-P_3-P_7-P_6-$P_8...$
Species 2: $...P_2$-P_9-P_3-P_7-P_6-P_8-$P_7...$ (with P_9 "restored")

By comparisons of this sort (taking into consideration likely deletions and other DNA-sequence perturbations), it was found that the most "parsimonious tree" for ribonuclease (an enzyme) largely agrees with biological common sense: A rat is close to a mouse; a buffalo is close to an ox; and a deer is close to a moose. On the other hand, a hippopotamus was found to be more closely related to a camel than to a pig, in direct conflict with opinion based on anatomical differences.

If ribonuclease sequencing shows that a hippopotamus is more closely related to a camel than to a pig, then data from other proteins should lead to the same conclusion. It turns out that α-crystalline (the protein of the lens in the eye) supports the ribonuclease-based tree, but unfortunately agreement is not always so consistent. Thus, according to the sequencing of one particular protein (myoglobin), birds are more closely related to mammals than to reptiles. But according to another protein (cyctochrome-c), the sequencing corresponds to conventional wisdom, namely that birds are closer to reptiles than to mammals.

Inferring taxonomic relationships from protein sequencing is beset with many other problems that complicate the life of the molecular evolutionist:

a) Not all amino acid disparities between two species carry equal weight because some of the 20 amino acid units comprising proteins resemble each other, while others are quite different in structure. Assume that units P_2 and P_8 have similar structures. Assume further that neither P_2 nor P_8 resembles P_5. A mutation that replaces P_2 with P_8 would, therefore, have a smaller effect upon the properties of the protein than would a mutation that replaces P_2 with P_5. This important distinction would escape consideration with a computer program that merely counts simple amino acid mismatches regardless of their effect on the properties of the protein.
b) As mentioned earlier, certain sections of an enzyme are especially important because they are directly involved in the actual reaction catalyzed by the enzyme. Other amino acids seem to be merely "along for the ride". A mutation in one of the latter amino acid units might have little impact upon the activity of the enzyme. It has been documented that functionally less important sections of a protein evolve faster than the important ones, again complicating the comparisons of sequences. Consider, for example, the proteins of three species. Only one vital sequence and a non-vital sequence within the proteins have been included:

	Vital Region	Non-Vital Region
Species 1	$\ldots P_1\text{-}P_8\text{-}P_5\text{-}P_9\text{-}P_6\text{-}P_6 \ldots$	$\ldots P_8\text{-}P_7\text{-}P_2\text{-}P_5\text{-}P_2\text{-}P_2 \ldots$
Species 2	$\ldots P_1\text{-}P_8\text{-}P_5\text{-}P_9\text{-}P_6\text{-}P_6$	$\ldots \mathbf{P}_5\text{-}P_7\text{-}P_2\text{-}\mathbf{P}_3\text{-}P_2\text{-}P_2 \ldots$
Species 3	$\ldots P_1\text{-}P_8\text{-}\mathbf{P}_4\text{-}P_9\text{-}P_6\text{-}P_6 \ldots\ldots$	$\ldots P_8\text{-}P_7\text{-}P_2\text{-}P_5\text{-}P_2\text{-}P_2 \ldots$

Mutational substitutions in Species 2 and Species 3 (relative to Species 1) have been darkened. The question, then, is whether

Species 1 is evolutionarily closer to Species 2 or Species 3. It can be seen that Species 2 differs from Species 1 by two substitutions in a non-vital region. Species 3 differs from Species 1 by only a single substitution, but it is in the vital region. If one counts only the total number of mismatches without regard to location, then one would conclude that Species 1 and 3 are more closely related than are Species 1 and 2. But since only Species 1 and 2 are invariant in the vital region, it might be concluded instead that these two are more closely related. Such are the perplexing problems and decisions inherent to molecular evolution.

c) A less fundamental limitation of molecular evolution relates to the fact that DNA and proteins are often not available from species that have long departed from the face of the earth. Construction of ancestral trees has, therefore, focused on living organisms.

How does Charles Darwin fit into all this? Most evolutionary trees based on molecular sequencing seem more or less consistent with those based on anatomy. Humans resemble chimpanzees more than dogs both in their appearance and in their genetics. Major surprises are rare. Molecular genetics provide comfort to those convinced of the continuous evolution of life, but molecular genetics (like anatomy) has provided no deep insight into the mechanism of evolutionary events. Natural selection is neither strongly supported nor discredited. But in fairness I should end with a notable success in DNA/evolution arena:

DNA testing led to the conclusion that whales had descended from land-dwelling "artiodactyls" (even-toed herbivores such as antelopes and hippos). This was surprising to many because at that time (1990s) most paleontologists agreed that whales had descended from a group of carnivorous Eocene mammals. In the year 2000, however, Philip Gingerich, who was working in Pakistan, discovered an anklebone from a four-legged whale dating back to 47 million years ago. The ankle bone resembled the corresponding bone in an artiodactyl, thereby confirming the contention of molecular biologists that whales and antelopes are related. It is particularly satisfying when evolution derives support from the convergence of multiple and independent lines of inquiry.

Thus, ends the survey of Darwin and his theory of natural selection. In the next chapter, I discuss another major figure in evolution, Jean Baptiste de Lamarck. Following this, the book takes a sharp turn. Darwinian and Lamarckian ideas are focused on perhaps the most remarkable evolutionary product of them all: Human intelligence. As will be seen, natural selection leaves us dissatisfied when it is applied to that remarkable organ-of-perfection housed in a "thin bone vault": The human brain.

It is recognized that this book addresses people of varied interests. Those who are primarily concerned with natural selection per se can stop here (with the author's hope that the discussion of this vast topic has been balanced and instructive). Those interested in the nature of human intelligence, and its possible origins, should read further now that a background in the principles of natural selection has been set forth.

Chapter 4

LAMARCK

Jean Baptiste de Lamarck (1774–1829) was by all counts a scientific pioneer. As an amateur biologist, he wrote his famous three-volume work "*Flore Française*" ("*French Flora*"). While discovering many new species of invertebrates, he laid down the foundation of our modern system of animal classification. But it was his "*Philosophie Zoologique*" (1809) and his "*Histoire Naturelle des Aminaux sans Vertebrates*" ("*Natural History of Animals without Backbones*") (1815) that elevated him to the status of a great evolutionist. Evolution is popularly attributed to Darwin, but this is grossly unfair to Lamarck who was the first to devote exclusively an entire book to the subject. True, Lamarck never documented his ideas with the exacting observational data as did Darwin, and Lamarck's "inheritance of acquired traits" (the main subject of this chapter) seemingly lacks the power and generality of Darwin's natural selection; yet the reality of evolution (including that of man) was championed by Lamarck half a century before Darwin. Lamarck was a giant on whose shoulders Darwin stood.

Not only has Lamarck received sparse credit for his accomplishments, he has been a victim of outright scorn up to the present day. C. H. Waddington expressed it well:

> Lamarck is the only major figure in the history of biology whose name has become, to all intents and purposes, a term of abuse. Most scientists' contributions are fated to be outgrown, but very

> few authors have written books which, two centuries later, are still rejected with an indignation so intense that the skeptic may suspect something akin to an uneasy conscious.

Suppression and denigration of Lamarck began with his French colleague (and Napoleon's protégé) M. le Baron Cuvier. It continued with Darwin himself who, along with his followers, effectively downgraded Lamarck's substantial contributions to evolutionary thought. Although in 1837 Darwin referred to Lamarck as a "lofty genius", Darwin's attitude in later years turned antagonistic. In a letter discussing books on evolution, Darwin wrote: "I do not know of any systematical ones, except Lamarck's, which is veritable rubbish." In another letter, Darwin wrote: "You often allude to Lamarck's work; I do not know what you think of it, but it appeared extremely poor; I got not a fact or idea from it." The latter claim is not quite accurate, witness two earlier quotes from Darwin's writings:

> I think there can be no doubt that use in our domestic animals has strengthened and enlarged parts, and disuse diminished them; and that such modifications are inherited.
>
> Species have been modified during a long course of descent. This has been effected chiefly through natural selection of numerous successive, slight, favorable variations, aided in an important manner by the inherited effect of use and disuse of parts.

In these last two quotes, Darwin adopts — unabashedly — the idea most commonly attributed to Lamarck, namely the "*inheritance of acquired traits*". Lamarck believed that a trait acquired in response to use and disuse could be passed on to future generations. The idea seemed to have intuitive appeal and was generally accepted in Darwin's times and for decades thereafter.

Examples of "inheritance of acquired traits" can be taken from Darwin himself. In the "*Descent of Man*", Darwin advised young women to learn as much as they can prior to starting families; the expectation was that this would help endow the future children with useful skills. In 1873, Darwin published an article in Nature, entitled "Inherited Instinct", which describes a dog's violent antipathy toward a butcher who had mistreated the animal. According to Darwin, the antagonism had been transmitted to at least two generations of the dog's offspring.

Neo-Darwinists, namely Darwinists armed with a knowledge of genetics that Darwin never possessed, completely discount the possibility of acquired traits being inheritable. The main reason for this lies in the difficulty of explaining how an acquired trait, such as antagonism to an unkind butcher, could be transmitted to the genes of the germ (reproductive) cells that, as everyone knows, are the seeds of inheritance. A mechanism whereby a change in the body leads to change in the gene apparatus has yet to be discovered. Until such a mechanism is proven to exist, Lamarck's theory will, in many quarters, suffer continued ridicule.

There is no point in citing additional unpleasant historical details. It is Lamarck's science, not the way he was maltreated, that merits discussion here. What are the implications of Lamarck's "inheritance of acquired traits", and are any aspects of his evolutionary mechanism relevant to modern biology?

Lamarck believed that the environment played a key role in evolution. Any alteration of conditions would result in new needs of an organism. Accordingly, the organism would, over many generations, meet its new needs via structural changes that, ultimately, become codified by heredity. Thus, Lamarck clearly voiced the concept, as did Darwin decades later, that organisms slowly adapt to their environment. What modern science finds so misguided and unacceptable is Lamarck's idea that adaptive traits acquired during the lifetime of an organism can become incorporated into the hereditary makeup. After all, cutting the tails off rats for tens of generations will not produce a tail-less newborn rat. And generations of circumcised boys have never produced a male child who is born already circumcised.

Actually, rat tails and circumcision are poor examples because Lamarck specifically dismissed modifications resulting from injury.

Instead, his focus was on modifications acquired by increased use or disuse of a bodily part in response to an environmental stimulus. For instance, Lamarck felt that the tiny ineffective eyes of moles arose from prolonged disuse of that organ underground. Continuously stretched skin between the digits of water birds led, he argued, to webbed feet. Of course, how an organism distinguishes between a "natural" stimulus and one that is imposed by injury was never fully clarified.

Lamarck believed that (a) generations of tanning in the tropical sun will lead directly to humans with dark skin; (b) a boy born to a family that had worked for generations as blacksmiths will inherit strong arm muscles; (c) a tough skin on the soles of feet is the birthright of a boy whose forefathers were barefoot peasants. One can appreciate the emotional appeal of Lamarck's route to evolutionary improvement. The mechanism is efficient and purposeful in a manner that macroevolutionary Darwinism, based on genetic accidents, certainly is not. There is yet another laudatory feature of Lamarckism: Lamarckism is more amendable to experimental testing than is Darwinism. Thus, one can modify an organism, generation after generation, by means of an environmental stimulus; one can then record whether or not the acquired modification at some point "sticks" in the absence of that same stimulus. Unfortunately, a negative result can always be blamed on the use of an insufficient number of generations, but nonetheless such experiments are worthwhile. I now present several ingenious attempts to test Lamarck's scholarly, if currently unaccepted, speculations on the origins of biodiversity.

In the 1920s, the great I. P. Pavlov, famous for conditioning dogs to salivate at the sound of a bell, became interested in the inheritance of acquired traits. He trained rats to run to a food source when he rang a bell. The first generation required an average of 300 trials to learn the exercise, the second generation required 100 trials, the third generation required 30 trials, and the fourth generation required only 10 trials. As any first-rate scientists would do, he attempted to repeat the experiments, but reported that they were "very complicated, uncertain, and moreover difficult to control". He concluded that: "The question of hereditary transmission of conditional reflexes and of hereditary facilitations of their acquirement must be left completely open."

As admitted by Pavlov, his experiments were not simple to execute. One must be careful not to inadvertently select the more intelligent rats in each generation. And one must also guard against distortions in the data caused by the researcher becoming more and more skillful in training the animals.

Prior to pupating, the willow-moth caterpillar crawls near the tip of a leaf and draws the leaf, beginning with the tip, around its body. The "leaf roll" is kept intact with a web. Fifty years ago a scientist by the name of Harry Schroeder wondered what would happen if the tip of the leaf onto which a caterpillar had positioned itself were removed. He found that the caterpillars solved the dilemma by rolling the leaf from the side rather than from the terminus. More interesting, Schroeder discovered that 4 out of 19 descendants of the side-rolling caterpillars also rolled from the side even when exposed to normal, uncut leaves. It appeared as if an acquired behavior had been inherited.

A further comment about the preceding experiment is necessary. I have extracted the basic information about Schroeder's work from secondary sources only. As such, assurances cannot be given here that the experiments were properly repeated, that controls were carried out, and that 4 out of 19 is in fact statistically significant. It is, therefore, impossible to present this work as anything but an intriguing preliminary observation. Under the circumstances, the best one can say is that the experiments should be repeated and the results double-checked under highly controlled conditions. If Schroeder's conclusions turn out to be valid, then this would, no doubt, have important implications for evolution. Unfortunately, carrying out tests of Lamarckian ideas today is neither popular nor, I presume, fundable by the granting agencies upon which all scientists depend. Other experiments described below suffer from the same problem. Too much science, to my taste, has died a death of uncertainty or obscurity more for reasons of fashion than for lack of inherent interest.

Here is an experiment that, being fond of animals (even rats), I would not repeat even if the funding for it were available. Rats were placed on slowly revolving turntables for periods of up to one-and-a-half years. When the poor animals were freed, their eyes and head constantly flicked in the direction in which they had been rotated. This flicking behavior also occurred, it was claimed by the investigator, in their non-rotated progeny as if the acquired trait had been inherited.

Work on ostrich calluses has been so widely quoted and misquoted that I felt obliged to consult the original literature (J. E. Duerden, *American Naturalist,* **54**, 289, 1920). A succinct account of Duerden's long article is given below:

a) Ostriches have an inherent ability to form calluses all over their skin — even in areas not normally subjected to pressure or friction such as the ankle region.
b) All ostriches develop calluses in the sternum (breastbone) region. It was at one time believed that these calluses were acquired from everyday activities (such as crouching on the ground to take dust baths whereby the bird rocks on its breast from side-to-side). But it was discovered, and this is critical, that ostrich embryos possess sternum calluses of the same form and nature as those found in adults. Thus, the sternum callous is an inherited trait transmitted from generation to generation.
c) In summary, Duerden discovered an inheritable trait that is identical to a trait that, clearly, would also be acquired rapidly from activities of the animal had the trait been absent from the very beginning of the animal's life. In other words, the gene for the sternum callous is not needed. Duerden wondered how a gene can be permanently retained, from an evolutionary standpoint, when the gene is unnecessary and is thus without apparent survival value. Three salient quotes from the paper bear repeating:

> An inherent power is transmitted and nothing is gained by transmitting the callosities themselves, since they are adaptations which could arise in the natural course as needed.

> It is legitimate to inquire whether a transmissible character is necessarily germinal as present-day teaching so consistently affirms.

> Natural selection has no bearing for [the callosities] are adaptive structures which the organism has the inherent power to produce as required.

Duerden concludes (guardedly) that the sternum callous may be an example of a transmissible but non-germinal trait and thus in conflict with neo-Darwinian precepts.

What is to one to make of all this? My instinctive tendency is to deny the assertion that a genetically-controlled sternum callus is "unneeded" and, consequently, of negligible survival value. To support my denial, I will resort to "evolutionary stories" in true Darwinian fashion. Here is a sample of possible responses to Duerden:

a) By inheriting calluses, the ostrich might acquire full protection in the sternum region earlier in life than would be possible environmentally via physical irritation of the skin in the course of bathing in dust. Survival value might be related to this time differential.
b) Genetically inherited calluses might be energetically less costly to form than calluses acquired by dust baths, etc. The gene for sternum calluses could have been preserved for this reason.
c) A callus present at birth might be only superficially identical to a physically induced callous. If the former were superior in some unknown way, the gene(s) for the embryonic callus could conceivably persist in the population.
d) The gene for the sternum callus might have a second unrelated purpose that is critical to the survival of the ostrich. Although in this rationale the inherited sternum callus is indeed unnecessary, it forms anyway because the gene controlling it has a more important additional role in the life of the ostrich.

Overall, I regard Duerden's controversial work as an unresolved issue. There is no compelling reason to believe Duerden's conclusion any more or any less than one of my four responses. And herein lies the great problem with evolutionary theories: Persuasive experimental or observational evidence is frustratingly hard to come by. Philosopher Karl Popper once asserted that Darwinism is experimentally untestable and is, therefore, a metaphysical construct. This is an extreme position. A more fair assessment would be that as yet no one has devised a definitive experiment proving one macroevolutionary model over any other. Models, in any case, are to be used, not believed. And Darwinism has thus far been our most useful model despite the fact that, for example, the appearance of a new genus has never been demonstrated experimentally even under conditions of high, artificially imposed mutation rates. Although evolution itself is a self-evident fact, the mechanism of evolution is best regarded as an unsolved problem. An open mind is, therefore, highly recommended (especially for the final pages of this book)!

The parathyroid gland helps maintain calcium levels in the blood. When the gland is removed (a "parathyroidectomy"), calcium levels decline. Clever experiments on the parathyroid, carried out by T. Fuji in 1978, are relevant to the Lamarck model:

a) The parathyroid was removed from pregnant rats. But their newborn offspring experienced little decline in serum calcium, during the first 24 hours of life, even though parathyroidectomies had been carried out on them at their birth. In other words, the parathyroid removal from the mother rat transiently protected the newborn from the effects of a similar operation.
b) Newborn rats from normal mothers were also subjected to parathyroidectomies at birth, but now the animals did indeed suffer a pronounced decline in calcium. Thus, none of these controls showed the "protection" evident in the previous experiment.

c) Brother and sister rats that had had a parathyroidectomized mother, but were allowed to keep their parathyroids, were mated. The progeny of such unions also produced young rats with a protective response upon having their parathyroids removed. The effect persisted for four generations.

The obvious implication is that an acquired trait, namely protection against parathyroid removal, is inheritable.

Or consider the experiments of Guyer and Smith in 1920. The experiments dealt with an acquired defect in rabbits characterized by an opaque lens and diminutive or sunken eyes. The defect virtually never appears in a colony of undisturbed rabbits. When, however, pregnant rabbits were subjected to an "anti-rabbit-eye extract" (prepared from chickens), about 18% of the young developed the pre-natal eye defect.

A series of matings involving many rabbits (defective and non-defective; siblings and non-siblings) led to a complicated "pedigree chart" extending over six generations. Suffice it to say that the appearance of the defect bore the general characteristics of a recessive gene. The simplest explanation for the data is that the externally administered defect had been laid down in the germline and passed on from generation to generation.

In the 1950s, C. H. Waddington exposed fruit-fly eggs to ether fumes for about 25 minutes. A few hatched flies were of the "bithorax type" having four instead of two wings. The abnormal flies were removed and mated, and the new egg batch again exposed to ether. As the procedure was repeated again and again, each successive generation showed a higher and higher proportion of the four-wing abnormality. After several generations (as few as eight), many eggs hatched into four-winged flies even in the absence of an ether treatment.

Originally, Waddington explained his results in terms of a Lamarckian inheritance of an acquired characteristics, but later he backtracked and offered the following more conventional picture: Rare genes, sensitive to ether shock, are scattered about the fruit-fly genome in various numbers. In the absence of ether, one is not even aware of their presence. In the presence of ether, however, they reveal their presence as an abnormality in the development of the thorax. When Waddington selected progeny showing the greatest effect of ether, he was actually selecting flies that carried the largest number of such ether-sensitive genes. With each generation, the genes were being concentrated until, ultimately, there were enough of them so that each fly developed abnormally even in the absence of ether. Waddington's research illustrates the difficulties that often arise in differentiating Darwinian and Lamarckian mechanisms.

M. Ho and her colleagues repeated the Waddington experiments in 1983, this time allowing all flies, normal and otherwise, to mate at random. Even though the abnormal four-wing flies were at a mating disadvantage, their numbers increased from 2% in the first generation to over 30% in the tenth. Upon return to normal conditions (no ether), the proportion gradually diminished over several generations.

Ho discovered that the tendency of the ether-treated eggs to produce four-wing flies was inherited through the mother. It was concluded that the ether modified the cytoplasm (i.e. the non-nuclear cellular material) rather than the genes themselves. This follows because cytoplasm is acquired mainly from the mother via the eggs. Moreover, ether-induced mutations of DNA are unlikely from a chemical point of view because ether is not a chemically reactive compound.

Although a cytoplasm-based theory of inheritance runs afoul of modern genetics, it is attractive in this particular case because egg cytoplasm should absorb ether, and be affected by it, far more readily than ether would mutate genes. Perhaps chemically-induced changes in cytoplasm are ultimately transmitted to the genes, so one cannot easily distinguish between cyctoplasmic and genetic effects. In any event, a transmittable cytoplasmic perturbation is decidedly Lamarckian in flavor; it signifies that an external factor can be inherited without initial involvement of DNA.

A listing of Lamarckian-type experiments must include those of E. J. Steele and R. M. Gorczynski in the early 1980s. To understand these experiments, one should be aware of P. Medawar's Nobel Prize work carried out in the fifties. Medawar showed that foreign cells injected into a newborn mouse will permit, later in the mouse's life, acceptance of a graft composed of the same foreign cells. Thus, Medawar was able to graft, onto a white mouse, a black patch from another mouse after first subjecting the white mouse, while newly born, to the black cells. The black foreign cells had obviously become non-immunogenic. In other words, early injection of black cells had caused the white mouse to become tolerant of a black-cell graft later in life.

Steele and Gorczynski found that 50% of white offspring from tolerant males were also tolerant to the black grafts even though the newly born white offspring had, unlike their father, never been exposed to black cells. The second generation of untreated whites was tolerant in 20–40% of the cases. It was concluded that acquired immunity to the black cells had been transferred to the germ line. In other words, the acquired tolerance is inherited. Attempts to repeat the work, prompted by the considerable publicity it elicited, failed in several laboratories. When parallel experiments of L. Bent did not confirm the Steele results, the British Broadcasting System in London arranged for a radio debate between the two. At one point in the broadcast, the argument became so heated that the microphones were turned off. The Steele and Gorczynski conclusions are still considered unproved.

Steele's explanation for the inherited acquired tolerance is described in his small book entitled "*Somatic Selection and Adaptive Evolution*". In it he proposes that mutations in body cells are spread to other body cells by viruses; eventually, the viruses transmit the mutations to the germ cells where they are passed on to later generations. Viruses are, accordingly, considered to be the key "messenger" in Lamarckian processes. The Steele mechanism can be summarized as: Environmental perturbation → mutation in body cells → transfer of mutational information to viruses → viral transmission of mutation to germ cells → transmission to next generation.

One reason that most Lamarckian experimentation lacks credibility today dates back to the early 20th century and the infamous Paul Kammerer affair. Kammerer, a brilliant young Viennese investigator, carried out experiments with two types of salamanders: (a) A black Alpine salamander that bears two offspring at a time and does so on land; (b) A spotted salamander that produces 10–50 offspring at a time in lowland waters. Kammerer switched the natural environments of the two salamander species. Thus, the spotted salamander was forced to live under Alpine conditions, and the black salamander in the lowlands. A remarkable thing then occurred: The black salamander produced multiple offspring, whereas the spotted salamander only two. The reversed reproductive patterns persisted in subsequent generations.

Experiments with the blind newt, Proteus, are even more startling. When Proteus was raised from birth in normal white light, no eyes developed. Kammerer discovered, however, that newts raised in red light did form eyes. The implication was clear. Proteus is normally blind not because it lacks the genes for eyes but, instead, because these genes are somehow suppressed. Environmental factors apparently induce the appearance of new organs — even as complicated as the eye — by negating whatever factors are suppressing the genes. Since the newts exposed to red light have newborn that are sightless, the eye trait does not persist, and the experiment does not bear directly on the Lamarckian mechanism. Yet the simple elegance of this work shows why Kammerer was a highly respected scientist.

Things, however, did not end well for Kammerer, nuptial pads being his undoing. Most male frogs possess rough dark-covered pads on their front feet so as to better grasp slippery females while copulating in water. A particular frog species, *Alytes obstetricans*, lacks such pads which are unneeded because the frog normally mates on land where the females are less slippery. When Kammerer forced *Alytes* to mate in water for several generations, they ostensibly developed nuptial pads in true Lamarckian style.

To make a long story short, it was ultimately discovered that the pads on *Alytes* may have been faked via the injection of India ink.

The fraud was never directly pinned on Kammerer, and there are speculations that one of his many political or scientific enemies was involved. In any event, Kammerer's work in its entirety was discredited by the incident, and Kammerer committed suicide at an early age.

The wonderful experiments of Kammerer — only one of which was directly tainted — have (to my knowledge) never been repeated. There are two main reasons for this. First, many experiments required long periods of tedious animal breeding incongruent with modern funded research where pressure for quick results often rules the day. Second, general antagonism toward Lamarck by the scientific community at large has, no doubt, discouraged further work in the area. Lamarckian experimentation is unfashionable and could well bring undesired controversy and funding problems to those who attempt it.

Lest anyone doubt the antagonism toward Lamarck, let me quote from Richard Dawkins' influential "*The Blind Watchmaker*":

> We can no more imagine acquired characteristics being inherited than we can imagine the following. A cake has one slice cut out of it. A description of the alteration is now fed back into the recipe, and .... the next cake baked according to the altered recipe comes out of the oven with one slice already neatly missing.

This quote, intended by Dawkins to ridicule Lamarck by analogy, merits a response:

a) Removing a slice of cake can be regarded as an "injury" of the type that Lamark pointedly disavowed. It is akin to removing rat tails. (I do not particularly like this argument because it relies on Lamarck's ad hoc exception to his theory, but I include it anyway for historical reasons.)
b) Suppose that the missing slice prompted (by an unspecified mechanism) an alteration of the recipe to include an additional line of instruction: "Remove a slice". In such an event, cakes would indeed end up with a slice missing. In effect, the additional

line has codified the "acquired trait" in the recipe for future generations.

c) If Dawkins' point is that no conventional recipe can ever produce, in the baking process per se, a cake with a slice missing, then such a cake is an absurdity whether one thinks in Darwinian or Lamarckian terms.

The reader will appreciate that my responses to the Dawkins' analogy are weak because, in part, his analogy itself is silly. I mention the quote mainly to illustrate again the scorn heaped upon poor Lamarck who, right or wrong, deserved a better treatment from later generations of scientists.

One final comment on the Dawkins quote: "He should have expressed his opinion in the first person singular rather than write: *We* can no more imagine acquired characteristics being inherited ..." (italics mine). I for one can at least imagine such a thing easily. I can imagine a bacterium suddenly acquiring the ability to produce human insulin, an ability that is passed on from generation to generation. I can also imagine a tomato suddenly acquiring a new inheritable capability to resist frost. It is easy to imagine such events because they actually exist through the wonders of molecular biology. The science of molecular biology is actually based on the inheritance of a laboratory-acquired DNA modifications. The key question, of course, is whether Nature can act in a similar fashion without human input. Is Nature capable of transmitting information from the soma (i.e. the body cells) to the germ cells (i.e. the egg and sperm)? And this brings up the subject of the so-called "*Weismann barrier*".

In the 1880s, a German biologist by the name of August Weismann established a principle that has since become known as the Weismann barrier. The idea was that information flows from the germ cells into the soma but never the reverse. In more modern terms, a mutation in the germ cells can affect the properties of the body cells (the latter, after all, arising from the former), but a mutation in the body cells will never affect the germ cells.

Although many neo-Darwinists still retain the notion of a one-way genome-to-soma pathway, molecular biologists have for all intents and purposes broken the Weismann barrier. Consider the following series of events: A mutation is induced in the blood-forming cells of an organism. Later, the mutated gene is removed and inserted into the DNA of a germ cell from the same organism. Thus, the progeny arising from the germ cell will possess the mutation and manifest the altered blood cells. In short, "information" has flowed, in an anti-Weismann manner, from soma-to-germ cells. Of course, the molecular biologist was the agent by which this information has been transmitted. One only need to postulate a non-humanoid mode of transmission, such as a virus or a mechanism yet to be discovered, to "imagine" an acquired characteristic being inherited.

A purist might refuse to grant historical credit to Lamark for the above blood-mutation scheme because it is genetically linked. (This is a common view but an ungenerous one because even Darwin, decades after Lamarck, had no concept of genetics.) A classical Lamarckian mechanism does not, of course, mention mutations as the initiating event but, instead, invokes an imposed environmental stimulus of some sort (e.g. active muscle use, the sun, exposure to water, etc.). In order to bring Lamarckism more into resonance with modern genetics ("*neo-Lamarckism*"), I must postulate the following: An external stimulus can, in certain instances, induce the production of a chemical messenger. The messenger affects the structure or expression of genetic material, thereby creating an inheritable trait responsive to the original stimulus.

I hasten to add that hard evidence for the postulate is rare. Some will consider it preposterous. But I should record the prediction that, eventually, evidence will accumulate that supports the mechanism, and, as a result, biology will be revolutionized. In the meantime, I will make use of the postulate in the second part of the book where I discuss the "thin bone vault", i.e. human intelligence. As will be demonstrated, the evolution of human intelligence simply does not make sense in the absence of a non-Darwinian construct.

In order to grasp the essence of neo-Lamarckian principles, it is useful now to cite again the example of Kammerer's nuptial pads. Recall that Kammerer claimed to have found that the *Alytes* land frog developed inheritable nuptial pads when forced to live in water. Assuming this was a *bona fide* observation, how might it be explained? (Note that in answering this question the emphasis will be on developing a mechanism for inheriting acquired traits, the particular example of nuptial pads simply being a contrived vehicle by which the mechanism is illustrated):

Assume that all the genes for nuptial pads in *Alytes* are present, but at least one of them is suppressed ("turned off"). Turned off genes are commonplace in biology. Assume further that this nuptial pad gene is turned off because an inhibitor (e.g. a protein bound to the gene) prevents the gene from expressing itself. Prolonged exposure of the frog to water apparently "turns on" the gene. This could occur in a number of ways. For example, the aqueous environment might induce production of a compound in the blood, an "*activator*", that displaces the inhibitor at the DNA site where the nuptial gene is located. Thus, the change in the frog's blood chemistry allows the gene to express itself during the embryological stages of the developing animal. Nuptial pads appear. The essence of the mechanism, in summary, lies in the chemical communication between the somatic tissue and the germ cells. Admittedly, this neo-Lamarckian "story" is contrived, vague, and fanciful, but probably no more so than those Darwinian "stories" which were discussed in detail earlier on. Actually, the neo-Lamarckian "story" has a molecular touch to it and, in this regard, possesses a measure of appeal.

The neo-Lamarckian construct incorporates a number of interesting elements. First and foremost, it presupposes that the genes for the nuptial pad are present (but hidden) in the *Alytes* frog. How, one may ask, did an apparently useless set of genes get there in the first place? Assume that in the course of evolution, the nuptial pads appeared on the scene among all frogs including *Alytes*. Since pads were unneeded in *Alytes*, one (or more) of the corresponding genes was inhibited but allowed to remain in the genome.

The required inhibitor could, conceivably, be the product of mutations that were retained by *Alytes* owing to their value in preventing the formation of a useless and costly organ. In other words, the nuptial pad genes were simply being carried along as an unexpressed and functionless bit of DNA baggage — unexpressed and functionless, but poised to make an appearance in a period of excessive rain, or a new terrestrial predator, or capture by a scientist such as Kammerer. At that point, *Alytes* must adapt to water in order to survive, and it is here where the hidden ("cryptic") genes, useless for countless eons, come to the rescue. The mechanism is fast, efficient, and responsive to a rapidly imposed need.

A neo-Lamarckian mechanism has a major advantage over a Darwinian mechanism in that the latter is excruciatingly slow. Darwinian evolution does not allow an organism to counter an abrupt change in the environment. Random mutational events, most of which are harmful or lethal, are unlikely to keep pace with abrupt environmental changes, a fact that has led to, and is currently causing, the extinction of many species. But neo-Lamarckian mechanisms could in principle save the day. The prevalence of neo-Lamarckian mechanisms, if present at all, is unknown. All I can say is that if I were designing an organism with maximum survivability, I would endow it with neo-Lamarckian capabilities.

Let us return to *Alytes* and confront a knotty problem with the neo-Lamarckian mechanism. When the egg and sperm of water-exposed frogs unite to ultimately form a new frog, only a tiny amount of water-induced activator is contributed to the union (the rest being distributed throughout the frogs). This might conceivably be sufficient to cause nuptial pad formation in the ensuing embryo. But in subsequent generations the activator would become so diluted that one can hardly imagine the nuptial pad trait persisting unless exposure to water was continued. The nuptial pad trait would be, in other words, highly unstable. Only if there were some means by which the activator could maintain its concentration is it likely that the trait could survive for even a few generations. In the absence of such an effect, the nuptial pad trait would disappear almost as quickly as it appeared once the frog returned to land. Although this too would benefit the frog, experience tells us that biology does not work that fast. Thus, in order to make the neo-Lamarckian construct a little less abrupt, one must postulate a continued production of activator even after exposure to water has been terminated. Perhaps activator formation resembles that of an antibody or, alternatively, the activator might be a self-reproducing peptide (the existence of which is known) that gradually disappears with time. It is pointless, however, to speculate further because the above is merely a neo-Lamarckian "story" of no greater substance than the neo-Darwinian "stories". Its main purpose has been to demonstrate that a neo-Lamarckian inheritance is, cake recipes and assertions to the contrary notwithstanding, an *imaginable* component of Nature.

Consider again the rat-tail argument against Lamarckism. Does the absence of tail-less rats after generations of surgical tail-removals really invalidate the above mechanism (or any of the many possible variations involving soma/germ cell communication)? Hardly. The experiment signifies only that rats have evolved so that there is, in this case, no "feedback" signal between the intact tail and the genes that control the development of the tail. The experiment is negative and anecdotal, and it has nothing to say about other biological attributes where soma/germ cell communication might indeed be operative. It is even possible that the disappearance of eyes in cave-dwelling fish might be related to neo-Lamarchian feedback. Another possible example of such communication follows:

Geneticist T. Sonneborn removed, by microsurgery, a piece of the cortex (outer surface) of *Paramecium*, a one-celled animal covered by cilia (small hairs). The researcher then reinserted the piece after first rotating it 180° from its original position. It was obvious that the piece had been rotated because the *Paramecium* now had a segment of cilia pointing in the "wrong" direction. Remarkably, the offspring of the *Paramecium* also had an inverted row of cilia. The acquired trait had ostensibly been inherited.

Lamarck stressed that "need" is the causal factor promoting the inheritance of acquired traits, but he can be forgiven for this error given that the science of genetics was unknown to him. Clearly, it is not need per se that induces a structural variation. Instead, the environment can, in certain instances, alter the expression of a gene (already present but, until that moment, inactive), and needs are met as a consequence. In other words, species might be poised to evolve rapidly according to certain demands by the environment. Unused sections of DNA represent an enormous repository for this evolutionary change. Should there be any doubt as to the presence of the vast hidden potential in the genome, one need only reflect again on the variation among dogs. Selective breeding from a few ancestral canines, with little use of mutational changes, has exposed the genes for dog varieties ranging from the terrier to the greyhound (the dachshund being an exception in that it apparently arose from a "chondrodystrophic dwarf mutation").

To summarize: I have proposed speculative mechanisms for biological change, but the purpose of doing so was not to persuade anyone that the environment, the soma, and the germ cells are inseparably linked. My goals were more modest. I simply wanted to illustrate that the environment, soma, and germ cells may be linked by complicated molecular mechanisms that have never been falsified and, consequently, must not *a priori* be discarded. In this light, it is erroneous (or at least premature) to write, as did Dawkins in his influential "*The Blind Watchmaker*", that "inheritance of acquired characteristics not only doesn't happen, it *couldn't* happen" (italics his).

There is an ironic twist to the complaints by Darwinists about Lamarkism. Darwinists have gone on the attack against Lamarkian ideas using arguments that are virtually identical to those levied by creationists against Darwinists! Two examples, again quoting from Dawkins, follow:

a) "Indeed, the vast majority of [acquired characteristics] are injuries. Obviously evolution is not going to proceed in the general direction of adaptive improvement if acquired characteristics are inherited indiscriminately."

Note the similarity here to creationists' criticism of Darwinism, namely that since random mutations are overwhelmingly harmful, they are an unlikely source of improvement.

b) "The eye has been a useful example before, so why not again? Think of all the intricately cooperating working parts: the lens with its clear transparency, its color correction and its correction for spherical distortion; the muscles that can instantly focus the lens on any target from a few inches to infinity; the iris diaphragm or "stepping down" mechanism, which fine-tunes the aperture of the eye continuously, like a camera with a built-in light-meter and fast special purpose computer; the retina with its 125 million color-coding photocells; the fine network of blood vessels that fuels every part of the machine; the even finer network of nerves — the equivalent of connecting wires and electronic chips. Hold all this fine-chiseled complexity in your mind, and ask yourself whether it could have been put together by the principle of use and disuse? The answer, it seems to me, is an obvious 'no'."

Sound familiar? If not, then go back and review the section on "Complexity" in which I defend Darwinism against an equivalent argument based on an incredulity that random mutations could ever produce something as intricate as the eye.

Three additional quotes add more to the point:

G. R. Taylor wrote: "We are in no position to dogmatize about the inward flow of information." The great paleontologist H. F. Osborn wrote in 1895 a statement that seems as true today: "If acquired variations are transmitted, there must be some unknown principle of heredity; if they are not, there must be some unknown factor in evolution." And E. Jablonka and M. J. Lamb wrote in their book "*Evolution in Four Dimensions*": "Contrary to long-accepted majority opinion, not all genetic variation is entirely random or blind; some of it may be regulated and partially directed. In more explicit terms, it may mean that there are Lamarckian mechanisms that allow "soft inheritance" — the inheritance of genomic changes induced by environmental factors." I like these quotes because they are honest admissions of our ignorance and the possibility that a new view of evolution is on the horizon.

I will end this section with one recent example of neo-Lamarckism. British and Swedish researchers (*Eur. J. Human Genetics,* **14**, 159, 2006) studied carefully documented historical records from Överkalix, an isolated community in northern Sweden. They found that the paternal grandfathers' food supply (i.e. whether or not there was a famine during their pre-puberty period) was closely linked to the mortality of the grandsons (but not the granddaughters). It was concluded that "the environment might be able to modify the germline ... imprints." In other words, a group of men were exposed to a diet (not normally considered, of course, an inheritable entity) that had a profound effect on the viability (disease resistance, etc.) of men living two generations downstream... a truly remarkable effect inconsistent with conventional genetics.

My intention in **Section 1** of the book was to present the strengths and weaknesses of natural selection in as unbiased a manner as possible. I concluded overall that natural selection is a viable component of evolution but likely not the only one. Problems and uncertainties with classical natural selection do indeed exist, but these must be tolerated until natural selection is supplemented or replaced with a better model. In any event, a familiarity with natural selection is a necessary backdrop to what will ensue in the remainder of this book: An analysis of human intelligence. This may seem like an abrupt change in emphasis, but in reality the topic flows logically from the fact that human intelligence is the most remarkable and baffling product of evolution. It is intelligence that makes us human. Without a grasp on the origins of human intelligence — of the "thin bone vault" — claims that the principles of evolution have already been mastered ring hollow.

Section 2

THE THIN BONE VAULT

"*God sleeps in minerals, awakens in plants, walks in animals, and thinks in man.*"

Sanskrit apothegm
(4th century BC)

"*If only we can find something in the biological world that Darwin cannot explain, perhaps life will have a meaning after all.*"

David Papineau

Chapter 5

INTRODUCTION

Intelligence! Although intelligence may be difficult to define, we all know how it pervades our everyday lives: Reading a newspaper, fixing a toy, driving a car, operating a word processor, caring for a pet, admiring a flower, solving a crossword puzzle, balancing a checkbook, playing a game of chess, filling out a tax form, planning a vacation, painting a picture, cooking a dinner, confessing a sin, reflecting on God, and (perhaps most importantly to survival) predicting future events. The source of this intelligence, the brain, makes us unique in the animal kingdom. Other animals may be "better" endowed with regard to ears, eyes, legs, or sexual organs, but none has the brain capacity remotely equivalent to that of human beings. How did this astounding intelligence — housed in the "thin bone vault" — come to be? This is the central problem of evolution (if not biology) and the subject of the remainder of the book.

Darwin himself recognized and puzzled over human intelligence. He wrote: "There can be no doubt that the difference between the mind of the lowest man and that of the highest animal is immense." Anthropologist Melvin Conner wrote of a 10-month child who, pointing to a butterfly, said: "Dat." "I suspect," wrote Conner, "what we are looking at is the most rudimentary form of what may be the key to being human: a sort of wonderment at the spectacle of the world, and to its apprehensibility by the mind: a focusing, for the purpose of elevation;

an intelligent waking dream. In that capacity...we find our greatest distinction." Paleoanthropologist Richard Leakey believes that humans are a fortuitous episode in the history of life, although he does confess to an almost religious "humility at the power of the human mind." Descartes said it the most simply: "I think, therefore I am".

The evolution of intelligence was recognized early on as a dilemma. Thus, Darwin's contemporary, Alfred Russell Wallace, remarked about the human brain: "An instrument has been developed in advance of the needs of its possessor." Wallace was expressing here the view that humans have developed a brain with far greater capacity than was required for survival, over the ages, as hunter-gatherers. If this opinion is correct, then we must search beyond Darwinism for an explanation for the mind. Darwinism is, after all, a mechanism by which biology and the environment remain more-or-less in phase; there is no provision for the appearance and maintenance of an expensive "super trait" that exceeds basic survival needs.

One cannot blame Wallace or anyone else for believing that early man acquired, and passed on to us, an inexplicably high level of intelligence. Who is not continuously amazed at the incredible human intellectual and technological accomplishments: Quantum mechanics, Hamlet, the silicon chip, the Ninth Symphony, the unraveling of the DNA code, the Sistine Chapel, the laboratory synthesis of chlorophyll, the laws of thermodynamics, the jet plane, and so on. Clearly, somewhere along the evolutionary path, man developed an ability to perform extremely complex and creative mental feats. How did this come about? What evolutionary forces gave rise to such a remarkable brain? In short, why are we so smart?

Civilization could develop only after formation of an effective means of food production: Agriculture. This first happened about 10 000 years ago in the Fertile Crescent of the Near East where Neolithic people became farmers. They used tools to hoe and reap, and they sharpened stones to cut down trees and process wood.

Village farming communities were formed. The dawn of civilization was upon us.

Thus, it took only 10 000 years to convert a savage (albeit a smart one) into a writer of the Divine Comedy; a chaser of animals into a moon-walker; a fire builder into a laser physicist; a cave-dweller into a worshiper at St. Peters; a fur-wearer into a manufacturer of nylon; a gatherer of wild fruit into a grower of genetically engineered plants; a spear-thrower into a launcher of missiles with atomic warheads; a star-gazer into a computer expert. Only 10 000 years or 500 generations were needed to effect these transformations. Even if 100 000 years were required for modern intelligence to evolve, this is still only a split second on the evolutionary time-scale spanning 3.5 billion years since life began. More will be said of the "time parameter" later, but at present it is only necessary to note that the great problem of evolution is not simply to explain how we became so intelligent; one must also explain how we seemingly became intelligent so quickly.

In tackling the intelligence problem, I have taken on an impossible task. Library shelves are filled with volume after volume on the brain and its capabilities. No matter what is written here, I can be charged with superficiality, a charge to which I must plead *nolo contendere*. Even worse, the information that I do manage to include is necessarily highly selected. Bias could appear, via this selection process, despite pages comprised solely of established facts. My only hope is that the facts have been selected wisely, and that my opinions and theorizing (important components later on in the book) are clearly differentiated as such from the more firmly accepted material.

In the next section, I will present definitions of such words as intelligence, mind, and consciousness. Since different people have different meanings for these words, operational definitions seem necessary. I then launch into a discussion of the magnitude of human mental capabilities. Are we merely "smart apes" or is there something truly unique within our "thin bone vault"? Finally, I inquire as to the origin of human intelligence. Was the intelligence required for successful prehistoric hunting-gathering on par with the intelligence directing modern activities? In other words, was Paleolithic man likely as intelligent as us? If so, then at what point might the trait have first appeared? What has been the role of language and culture in the nurturing of our intelligence? Were three key parameters governing the magnitude of evolutionary progress (time span, population size, and mutation rate) sufficiently extensive to allow the notoriously slow Darwinian mechanism to manifest itself, or must we search elsewhere for an explanation of intelligence?

There is no point in being coy by withholding until later in the book a main conclusion of my analysis. Thus, I will ultimately argue that human beings possess a level of intelligence that cannot be reasonably explained by a strict Darwinian construct. Although I consider myself a Darwinist of sorts, and in a previous chapter I have defended natural selection against a host of criticisms, I find myself unable to do so when it comes to the intelligence issue. Intelligence arises from an intricate and highly evolved neural network that is extremely costly to operate (the brain consuming roughly 20% of the body's intake of oxygen although comprising only 2% of the body weight). It allows human beings to routinely perform remarkable mental feats that surpass any possible survival needs of recent ancestors living, say, only 20 000 years ago. If these people owned a modern intelligence "in advance of the needs of its possessor" (in the words of Wallace), then a Darwinian explanation for the human brain fails, and fails badly, because, as we know, natural selection has no foresight, no plan. If, on the other hand, our complex and costly intelligence evolved only in modern times, then Darwinism must be sidestepped with equal determination; organs of perfection do not, according to natural selection, evolve in the course of a few thousand years. Either way, we face a serious predicament. The present book on the "thin bone vault" attempts to provide a more acceptable rationale for the human mind. My intention is not to replace natural selection but to expand it and, thereby, to enhance our understanding of Nature.

Intelligence may not be easily defined, or its evolution understood, but there is no doubt that the brain is the seat of our intelligence. Although this book need not devote itself to details of brain physiology and anatomy, a brief description of the brain will be useful. Reptiles have an "old" or "primitive" brain that regulates blood pressure, sex, emotions, and movement. Humans and other mammals also possess such an "old" brain but, in addition, they are endowed with a *neocortex*: A soft, six-layered, 2 millimeter thick sheet of neural tissue that covers most of the "old" brain. Most of what we regard as intelligence (perception, language, imagination, etc.) is housed in our neocortex, with other brain parts (the thalmus, hippocampus, etc.) also contributing. It has been estimated that the human neocortex contains about thirty billion (30 000 000 000) nerve cells (called neurons). In simple terms, we are smarter than other mammals primarily because we have a larger neocortex.

Extreme flexibility — a capacity to absorb and adapt to innumerable environmental stimuli — is perhaps the key feature of the human brain. The human brain can, according to the particular environmental exposure, learn one or more spoken languages, various written languages, and musical, mathematical, and computer languages. Neocortical memory in turn allows the prediction of patterns, probably the most important component of our intelligence. Consider, for example, the Nobel Prize-winning discovery of penicillin by Alexander Fleming. Let me construct a possible thought process that might well have passed through Fleming's neocortex: "My bacteria culture is normally cloudy, but the culture has been contaminated by a small circle of fungus growth, and the area surrounding the fungus colony is absolutely clear (*observation*). Last week I recall seeing this same type of clear area surrounding the mold on a culture (*memory*). This seems to be a recurring and reproducible effect (*pattern recognition*). I recall that whenever I place an antibacterial drug on my plates, the cloudy bacteria colony is killed, and the culture clears up (*memory*). This suggests that the mold is producing a bacteria-killing substance (*pattern recognition*). I will isolate the substance and call it penicillin (*prediction*)." It is not an exaggeration, therefore, to state that we owe penicillin to Fleming's neocortex. More generally, the story of intelligence is the story of the neocortex. Evolution expanded this incredible organ rather quickly in humans, perhaps over the past couple of million of years, but we do not know how or why.

Chapter 6

DEFINITION OF INTELLIGENCE

Part 1: A Multifaceted Trait

Many people think of intelligence in terms of "IQ", but in actual fact intelligence is far too complicated to be properly embodied by a single number (which is not to say that IQ is devoid of utility). From my standpoint, IQ is best defined as "whatever an IQ test measures as determined by a group of test writers". Since this is neither very informative nor satisfying for our purposes, "IQ" will never be mentioned in the ensuing discussion of human intelligence and its evolutionary origins.

Most teachers of chemistry, such as myself, have encountered intelligent students who have received high grades in challenging or abstract subjects such as Russian, economics, and philosophy. Yet these same students complain bitterly about problems in mastering their chemistry courses. The reverse also occurs, suggesting that poor teaching by the chemistry professors is not the only source of trouble. A more plausible explanation is that there exist multiple intelligences. Each person is good at certain types of mental feats but less so in others. Pablo Picasso, for example, is unlikely to have made (one surmises) a leading mathematician or Albert Einstein a preeminent artist. Success with one mental facility often has little predictive power for success with another. If, therefore, intelligence is not merely a single attribute that is possessed by every human being to a greater or lesser extent (as is, for example, skin pigmentation), then it behooves us to inquire into the components of the trait. In order to do this, I will define intelligence in terms of the classification given in Howard Garner's book entitled "*Frames of Mind*". A few pages devoted to this definition of intelligence seem important here because too often a reader is exposed to terms such as "intelligence", "mind", "conscious", etc. without ever being handed a firm grip on exactly what is meant by the author.

Gardner subdivides intelligence (somewhat artificially but usefully) into the following more-or-less autonomous subsets: linguistic intelligence, musical intelligence, logical-mathematical intelligence, spatial intelligence, bodily-kinesthetic intelligence, and the personal intelligence. Critics of this classification feel that "intelligence" should not be confused with various "mental talents" but should, instead, be reserved for an all-encompassing mental capacity. There is no need here to become embroiled in matters of semantics. I do feel, however, that a brief survey of "intelligence types" or "mental talents" (whichever term one prefers) provides an excellent opportunity to portray the amazing breadth of the human mind. Ultimately, of course, it will be necessary to ask from where our mental prowess originated. At that point, the previous chapters on evolutionary theory will reassert themselves.

Part 2: Linguistic Intelligence

Language, the vehicle of explanation, persuasion, desire, reflection, recall, and prediction, is perhaps the most universally shared component of intelligence. Barring disease, defect, or injury, all humans have the ability to express themselves via language. And human capabilities in the realm of language are nothing short of astounding, the people of Switzerland providing a prime example. According to 1998 statistics, 1 420 597 Swiss (in a country of 6 873 687) speak at least two national languages. Among these, 242 432 speak three languages fluently: German, French, and Italian. Almost 13 000 people speak four or more languages. Thus, the ability to speak multiple languages is not a rare and isolated talent found in the occasional genius; it is commonplace in Europe and elsewhere. No doubt a greater fraction of the United States population would also be polyglots were there suitable motivation and a better educational system.

Professional basketball star Dikembe Mutombo was born in Kinshasa, capital city of Congo, the seventh of ten children. Mutombo is fluent in nine languages including five African languages. Although this is anecdotal evidence for remarkable language skills among humans, it is nonetheless impressive.

It might be argued that the learning of several languages is simplified by groups of languages (e.g. the Romance languages) being interrelated. Other languages (e.g. English) contain significant elements of several languages. While this is all true, one must not invoke linguistic interrelationships to dismiss or even minimize the extraordinary intelligence manifested by multilingual individuals. Consider the words "cat" and "butterfly" and the phrase "I do not know":

English:	cat	butterfly	I do not know.
German:	*katz*	*schmetterlink*	*Ich weisse nicht.*
French:	*chat*	*papillon*	*Je ne sais pas.*
Spanish:	*gato*	*mariposa*	*No se.*

These words and phrases are sufficiently different from language to language that knowing only one of them does not automatically guarantee knowledge or even recognition of the others by inference. For example, a French person learning Spanish must "work" to learn the word "gato" despite the fact that French and Spanish are both Romance languages. And the correct gender, varying with the language for any particular noun, must also be memorized, not to mention the particular rules of grammar when the nouns are incorporated into full sentences. The point here is to belabor the obvious, namely that humans possess an amazing linguistic capacity — a capacity characterized by the ability to learn vocabulary and grammar far exceeding the contents of any single modern language.

Perhaps readers will feel that I have been unfair in my selection of "cat", "butterfly", and "I do not know" to illustrate a point. Granted, English is a Germanic language and contains hundreds of words having the same meaning as their close German cognates. The English "thank" and "help", for example, are given by "*danken*" and "*helfen*" in German. Although such similarities are helpful to the language student, one must be careful here. The English "night" and German "*nacht*" are cognates (i.e. related by descent), and they sound alike and have identical meanings. Yet this overlap is not all that instructive since the unrelated German word "*nicht*" (meaning "not") sounds even more like "night". The German word "*rot*" means "red" and has nothing to do with the English "rot". The German "*baum*" is a cognate of "beam" but means "tree". "*Luft*" means "air" although it is a cognate of the English "loft". "*Tier*" means "animals" although it is a cognate of the similar-sounding "deer". "*Raum*" means "space", not room. "*Fahren*", related to "fare" (as in fare thee well), means "to travel". "*Sterben*" means "to die" although it is a cognate of "starve". The bottom line is that cognates both help and confuse. The bottom line here is that the person who speaks both English and German does so not simply because English and German have certain similarities, but because the human brain can process an immense quantity of linguistic information.

A key question confronts us immediately: Did prehistoric man possess a linguistic intelligence equivalent to that of modern man? Or, asked another way: Is it possible that prehistoric man had the capacity to learn the vocabulary and grammar equivalent to two or more modern languages? If the answer is "yes", then we must ask: "To what purpose?" "Why would prehistoric man require multilingual skills as found in, say, Swiss graduate students?" If, on the other hand, prehistoric man did not have modern language intelligence, then we must ask: "How did such a capacity develop so precipitously, i.e. over a trivial time-span (evolutionarily speaking) of a few tens of thousands of years?" With either "yes" or "no" in answer to the question of prehistoric language intelligence, one faces a puzzle of major proportions.

Early man obviously lacked an educational system in which multiple languages were taught. But it is necessary to forestall any claim that our current facility with multiple languages is due solely to education ("culture"). Although education is obviously a great help in learning languages, even the best imaginable education would be useless if the human brain did not have the capacity to absorb it. Education works well only because humans are *educable*, and the latter can be attributed solely to the contents of the "thin bone vault".

Being well versed in Darwinian "stories" from a previous chapter, we will have no trouble devising a story "explaining" why prehistoric man might have possessed an inherent ability to speak the equivalent of multiple modern languages. As hunter-gatherers, these people lived in small bands that, on occasion, had to communicate with each other (perhaps to exchange information on where game animals had migrated, etc.). Consequently, there was survival value in a band learning the particular mumblings of other bands in the region. By this means, man evolved a capacity to comprehend a huge variety of words. The ability was further enhanced by a tendency (common primarily in primitive cultures) to adopt multiple words for single items like "snow" and "mosquito" according to their specific properties. For example, there might have been separate names for freshly fallen snow, icy snow, powdery snow, and drifted snow, all adding to early man's vocabulary. Thus, one concludes, primitive man probably had linguistic skills every bit as impressive as those of modern man.

Is this story plausible? I, for one, find it decidedly specious. Encounters among small bands scattered across vast terrain were probably not that frequent. (The point will be amplified in a later chapter on Population). Even if two bands did happen to meet, they were far more likely to kill each other, it would seem, than to exchange pleasantries (and language skills) over a campfire. And since, as we shall soon see, elements of language are so hard-wired in our neural network, early man would have had to experience, according to Darwin, survival advantage by using appropriate grammar. The end result of this process, in a trivial number of generations (evolutionarily speaking), is a Swiss student who can speak grammatically correct Swiss German, German, French, Italian, and English. It simply defies credulity. Yet, and this is an annoying "yet", lack of credulity does not constitute a disproof per se. Is it unlikely that early man's survival depended upon exploiting an innate linguistic brain power equivalent to what is in possession by mankind today? Yes, it is unlikely. Is it untrue? No one can say. Owing to this uncertainty, I will focus below less on speech (whose evolutionary details are mainly speculative) but more on writing (whose history is on reasonably firm ground). For example, we know when writing first began and how writing changed over time once that milestone was reached. As will be shown, we also know that writing became hard-wired in the brain.

When learning how to drive a car, we require another person to instruct us on how to accomplish the task. No "sense of driving" exists within us that would allow operation of a vehicle in the absence of instruction or demonstration. Even an act as trivial as inserting a key in order to start the engine must be taught. The situation with learning a language is quite different. There is increasing evidence that much of our linguistic intelligence is innate (i.e. hard-wired in our brains). Thus, a child picks up syntax and grammar without rules ever being explicitly taught. Naturally, a child also learns by imitating adults, but the acquisition of language clearly runs much deeper than the simple ability to mimic. In the remainder of this section, I will cite recent evidence that supports the notion of an inborn capacity for language that, mysteriously, appeared in humans at a level unique to the animal world. This capacity resides in special regions of the brain, and thus in our genetic makeup, and thus in our evolutionary endowment. No theory of evolution can be considered complete until it explains how it happened that the ability to decode language became entrenched in our nervous system.

Although ethical considerations preclude experimentation with the human mind, the study of patients with brain pathology (arising from injury, stroke, or congenital defects) has allowed important conclusions with regard to language and the brain. It has been observed, for example, that among highly aphasic patients (i.e. people who have lost their ability to use and understand the spoken language) many are able to perform well on cognitive tasks such as art, music, and math. It is as if language is a semi-autonomous component of intelligence not linked (at least directly) to other aspects of brain function.

Psychologist Alfonso Caramazza believes that speaking and reading the same word involves two different centers of the brain. Evidence for this was provided by two women who had suffered strokes in different locations of their brains' left hemisphere. Although both had trouble with verbs, the problems were in striking contrast: Patient A could write verbs but not speak them, whereas Patient B could speak verbs but not write them. Neither had any trouble writing or speaking nouns. For example, the following sentence was read aloud to the women: "There is a crack in the window." Both women were able to write correctly the word "crack" (nouns, as mentioned, presenting them no difficulty). But only Patient A was able to write the word "crack" when the sentence, "Don't crack the nuts in here", was read aloud to her. On the other hand, only Patient B was able to read aloud the word "crack" when the sentence, "Don't crack the nuts in here", was given to her in written form.

A word of caution is necessary when entering the world of brain pathology and its relationship to language. Such research is relatively new on the scene, and many reports are still preliminary and anecdotal. The conclusions are complicated by the fact that damage from two strokes is seldom exactly the same; that a particular brain function may be controlled by more than one site. Thus, the effects of brain anomalies on linguistics abilities, although a fascinating field, should for the moment be viewed guardedly.

With this caveat in mind, one can conclude from Caramazza's studies that there are separate compartments in the brain for processing nouns and verbs. The results also suggest distinct centers for speaking and writing. As a matter of fact, there is even evidence now that points to language centers in the brain that deal with word-endings such as "-ed" and "-ing". Thus, some of Caramazza's patients could understand the meaning of "stop" but not the word "stopping".

What does all this have to do with a book devoted to evolution? It has everything to do with evolution. Consider only the fact that human brains seem to have writing centers distinct from speaking centers. The evolutionary implications of this astonishing feature of our linguistic intelligence can best be appreciated in the context of human history. I will therefore digress a moment and discuss briefly the invention of writing — surely one of the greatest achievements of mankind.

The so-called "logographic writing" constituted a major advance in the development of writing; for the first time symbols began to embody the sounds of spoken words. Suppose, for example, that the mark ‡ is used to symbolize the word "bee". The symbol could, however, also serve as the first syllable in the words "beyond" and "become" as well as the final syllable in the word "baby". In this manner symbol and sound become entwined. Such a logographic writing system developed independently among the ancient Egyptians (as hieroglyphics), the ancient Chinese, and the Sumerians (as cuneiform).

Cuneiform was first used by the Sumarians at about 3200 BC in Mesopotamia (what is now southern Iraq). The writing was produced by pressing the sharp edge of a reed into wet clay, producing its characteristic wedge-shaped symbols. Importantly, there are no traces of writing, or forerunners of writing, in archaeological strata below the one in which the first cuneiform tablets have been found. Either there were no predecessors of cuneiform or else all evidence of them has vanished. Cuneiform is composed of roughly 600 symbols, about 75% of which represent words and the remainder represent syllables. Syllables could be used alone or with other syllabic signs to spell out a word phonetically.

Sumarians used combinations of two-signs to make a new word. For example, the sign for "woman" adjacent to the sign for "foreign country" signified "slave girl". The word for "wood" was used in concert with objects made of wood (a type of "ideogram" that reduced the ambiguity of their symbols). Most interestingly, certain word-signs lost their association with words and came to represent only the sound of the word. For example, the word for "arrow" (pronounced "ti") was given its own word-symbol. The word for "life" (redundantly also pronounced "ti") originally lacked a symbol, but, owing to the phonetics of the situation, the symbol for "arrow" was eventually adopted as the symbol for "life" as well. And, as mentioned, symbols were also extended to syllables comprising multisyllabic words.

Although logographs advanced the cause of writing, they had drawbacks. Most notably, a huge number of precisely drawn symbols are required in any logographic system (Chinese utilizing about 5000 of them). This problem led ultimately to the development of the alphabet of which there are about 50 in use today. An alphabet employs symbols, called letters, to represent elementary units of sound (e.g. the symbols "p" and "b" refer to the initial sounds in "pin" and "bin", respectively). Of course, we still retain various symbols to depict certain items (e.g. numbers, +, and &).

The history of the alphabet is one of the most fascinating chapters in human intellectual development, but for the purposes of this book it is necessary to mention only one more fact: All alphabets of today are believed to have descended from the Proto-Semitic script devised in the Syria–Palestine region between 2000 and 1500 BC. The dates for the first appearance of alphabetic writing, along with other dates for comparison, are given in the table below.

	First Appearance (years ago)
Mammals	200 000 000
Homo erectus	1 800 000
Neanderthal	500 000
Cave paintings	50 000
Homo sapiens (modern man)	30 000
Cuneiform writing	5000
Alphabet	4000
Birth of Christ	2000

As is obvious from the table, the appearance of writing is, by any measure, a "modern" event when placed in the context of human evolution spanning millions of years. Note that it is difficult to relate actual writing to those wonderful prehistoric cave paintings of southern France, drawn some tens of thousands of years ago, because artwork is a completely different intellectual exercise than symbolic representation of words.

Let us now return to the observation that initiated this digression into the history of writing: Human brains seem to have a writing center distinct from a speaking center. Stated more precisely, in the process of learning to write and speak, connections in different parts of the brain become wired to handle the chores. In order to explain how such an intricate mental capability came into existence, two possibilities present themselves:

a) The writing centers (i.e. the sections of our neural network that become appropriately "wired" in response to writing needs) may have made its appearance when writing first became a human activity (about 5000 years ago). If this is true, then Darwinism provides no satisfactory explanation, because the theory is grounded upon a gradual evolution of complex traits over vast periods of time. Darwinism cannot accommodate a precipitous appearance of a complicated neural network which, among other linguistic services, directs the writing of verbs in one location, nouns in another. But let us suppose that, preposterous as it may seem, that the mental circuitry for writing was indeed created in one fell swoop 5000 years ago, whereupon humans in the Middle East began primitive writing. If this were true, then how did the incredible capacity to read and write get distributed to each and every human being on earth within the course of only 250 generations? What possible Darwinian reproductive advantage might have caused the trait to spread universally at lightning speed (evolutionarily speaking) among a comparatively sparse population composed mainly of illiterates?
b) Alternatively, the ability to handle writing may have been firmly in place, say, 200 000 years ago (i.e. among the Neanderthals),

> but the skill was never actually put to use until the Sumarians. If this were true, then one must wonder why prehistoric man never exploited this wonderful innate ability to communicate (although one can always seek refuge here in that all-encompassing word "culture"). More to the point, why did the trait persist in the first place since, being unused for tens of thousands of years, there was no Darwinian advantage in possessing it? As a matter of fact, since maintaining brain tissue is, energy-wise, a particularly costly enterprise for an organism, one can far better visualize a major Darwinian ***disadvantage*** to any unused brain center. Cave-dwelling fish beneficially lose their unused eyes, so why should unused mental capacity in humans persist?

Parenthetic mention should be made here of the ASPM (abnormal spindle-like, microcephaly-associated) gene that is believed to have arisen about 5800 years ago and associated with enhanced brain development. Only 30% of the world's population have this gene, although it has been argued that evolutionary pressures are causing it to increase in the population at an unprecedented rate. At least two responses to this development seem justified: (a) No one seriously believes that a single gene, or even a small group of genes, can explain the difference in mental skills between humans and non-human primates. (b) Modern advances in public health have made most of us in the developed world "fit" in the sense being able to reach the reproductive years necessary to propagate our genes. Whatever is causing the spread of the ASPM gene, it is not natural selection.

In summary, if the complex writing centers in the human brain became available only 5000 years ago, then one must dismiss Darwin's gradualism. And if they became available long before that time, then one must dismiss Darwin's stipulation of reproductive advantage. Either way, Darwin appears frustratingly deficient. It seems highly improbable that our writing ability burst forth suddenly only 5000 years ago. In all likelihood, therefore, we have been owners of an amazing neural network that was sufficiently flexible ("*plastic*") to accommodate writing centers when the need arose. This does not solve the evolution problem, however. One must explain how a brain evolved that had the capacity and flexibility to face a challenge long before we ever had to face this challenge.

The mystery regarding the origin of the brain's writing centers presupposes that such centers, fully formed only after birth but formed nonetheless, do indeed exist. Since I have mentioned thus far only one set of experiments verifying the presence of writing centers (i.e. those dealing with the verb-writing problem), perhaps additional evidence is in order. Consider now the example of a condition called "*dysgraphia*".

Dysgraphia, a disorder or writing and spelling, often results from stroke, tumor, or trauma in the left parieto-occipital cortex of the brain. The condition assumes many forms. Thus, some patients write, in response to dictation, only what they hear (e.g. "*yot*" for "yacht"). Others write words that have only approximate similarity in meaning to what they hear (e.g. "moon" for "star" and "bun" for "cake"). Still others, unsure of how to spell a word, omit uncertain letters (e.g. "*tur y*" for "turkey"). A case has been reported in which a patient always deleted, and left spaces for, the vowels (e.g. "*b l gn*" for "bologna"). There is, of course, no apparent reason why vowels should be more difficult to recall or execute than consonants. Another patient, who suffered from a left-hemisphere infarct, did not omit the vowels but, instead, substituted or transposed them (e.g. "*caro*" for "*cora*", the Italian word for "dear"). Such vowel errors appeared in written and oral spelling, typing, and delayed copying. Single dictated vowels, incidentally, were accurately written down by the patient.

One might conclude (tentatively, as much more work in the area is needed) that human brains create a center that processes the writing of vowels. Pathology in this center leads to omission or transposition of vowels during attempts to write dictated words. Consonants are not affected; speech is not affected; only written vowels are recorded abnormally. The presence of a brain center for written vowels has far-reaching implications in human evolution.

To grasp the importance of the discovery, let us go back to a point in prehistorical times (whatever time period one prefers) when a neural capacity to create a vowel center did not exist. According to the Darwinian mechanism, there must have been a mutation (or, more likely, a fortuitous sequence of them) that imparted to some individual an ability to write vowels. Somehow or the other, this individual's brain received a major rewiring or expansion to accommodate written vowels. Now, according to Darwin, the vowel trait must have had considerable survival or reproductive value since no healthy human being today is without the trait. Survival value? What Darwinian story could possibly explain away the survival value of a center for written vowels? If some Cro-Magnon or Neanderthal or early primate (take your pick) appeared suddenly on the scene with an ability to write vowels, what would he or she do with it? There existed no alphabet with which to exploit the ability, let alone anyone to read the primitive scribblings had our "genius" attempted to communicate through writing. Survival in our extensive era of hunting-gathering (prior to 10 000 years ago) would, it seems, have depended more on wielding a spear rather than recording a written vowel.

While on the subject of vowels, mention should be made of the fascinating work of T. Tsunoda, a specialist in speech and hearing disorders in Tokyo. Japanese is a language that, unlike English, can express thoughts with sentences composed largely or entirely of vowels. For example, the sentence "A love-hungry man who worries about hunger hides his old age and chases love" translates to "*Ue o ui, oi o ooi, ai o ou, aiueo*". Now Dr. Tsunoda noticed that some Japanese patients who had suffered a stroke in the left hemisphere lost their ability to express the single sound "ah". In sharp contrast, non-Japanese who had moved to Tokyo as adults (including Americans, Europeans, Koreans, and Chinese) process certain single vowel sounds primarily on the right side. Even more interesting, Americans and Europeans who had been raised in Japan since an early age were "left-brained" just like the native Japanese. These results point to what we all know: That culture is an important determinant of language. One must not conclude, however, that genetics is superfluous. It is the genes that ultimately impart to the brain an ability to handle language, to form grammar centers when cultural factors so demand, e.g. to allow different locations of the brain to "connect" according to the properties of the language.

Few people believe that a brain capable of forming verb and vowel centers, as well as the host of other language faculties, made their appearance in recent times. Anthony Smith is adamant on the subject:

> The overriding fact, never to be forgotten, is that our near antecedents have all been similarly equipped mentally. So far as is known, simple hunter-gatherers, early cave-man, cave artist, Neolithic agriculturist, Sumerian tribe, and modern office commuter have all possessed the self-same brain.

When asked why or how such a magnificent development of brain power could have occurred among early, primitive, and scattered tribes of men, the famous anthropologist Richard Leakey said: "I haven't got the foggiest notion." Darwin was equally helpless to explain the contents of the "thin bone vault".

Before ending this section on language intelligence, I would like to give one final proof that a center in the brain devoted to reading/writing centers is formed in the human brain (a center that, as mentioned, has been put into actual use for no more than 5000 years). *Dyslexia* is a condition in which otherwise normally intelligent people have difficulty reading and writing. Since the syndrome tends to appear in families, there is a genetic component to it. Among the most common problems among dyslexics, one can cite the following: (a) Confusion between letters similar in shape (e.g. *d* and *b*; *u* an *n*); (b) reversals (e.g. "saw" for "was"); (c) transposals (e.g. "left" for "felt"; "auction" for "caution"); (d) repetition of words; (e) misplaced punctuation and capital letters; (f) difficulty in keeping place on a line or in moving from the end of a line to the start of the next; (g) letter fusion (i.e. letters not fully separated); and (h) confusion of letters with similar sounds (e.g. *v* and *f*).

Although the exact cause of dyslexia is unknown, A. M. Galaburda and T. L. Kemper were able to relate the condition to a disruption in the normal neuronal architecture of the brain. Upon examining the brain of a dyslexic man who had died of internal injuries after a fall, they found a disorganization within the language area of the left hemisphere. "The layers were scrambled and whirled with primitive, larger cells in this part of the brain." Thus, dyslexic errors in reading/writing were directly attributable to structural anomalies in the brain. The point here is to affirm (should there still be any doubt) that the human ability to write is a complex, universal, and innate talent that was, in all likelihood, firmly in place long before any use of it was made 5000 years ago. The potential for writing is firmly embedded in the brain at birth with cultural stimulation (education) allowing the potential to manifest itself. This assertion conflicts with the very heart and soul of natural selection.

Am I not being inconsistent by defending Darwinism in the "evolution of the eye" question (as I did in a previous chapter) while simultaneously claiming that Darwinism is powerless to explain the writing centers in the brain? Actually, there is no inconsistency because the cases are entirely different. There is a long line of increasingly complex eyes in Nature; the survival value of sight is obvious. Although we might not understand how the human eye evolved, it is easy to imagine that natural selection over a vast time span might have played a role. Not so with the sophisticated and specialized writing centers in the brain. These centers have been in direct use for only 5000 years. Prior to that, the centers were unused and therefore of zero survival value. Yet every human on earth has one. While the eye may be compatible with a natural selection mechanism, the writing center is certainly not. Something profound is clearly missing in natural selection.

I am hardly the first to point out the difficulties in explaining the human brain by natural selection. Darwin and Wallace, and innumerable scientists after them, were also concerned about the problem. For the most part, the problem was relegated to a "puzzle" and put on the back burner. It is one purpose of this book to reinstate the problem and to give it a fair hearing rather than to simply dismiss it as a "puzzle".

In a rather contrived attempt to "explain" the existence of a vowel center in strictly Darwinian terms, I have concocted a Darwinian-style "story": Once upon a time, perhaps 100 000 years ago, an ancestor appeared on the scene with a genetic modification that endowed him or her with a unique neurological capability: The mental wherewithal to read and write. Since reading and writing were obviously not part of the culture, the trait would have gone unnoticed were it not for the fact that the new mental prowess also assisted the ancestor to travel at night via guidance from the stars. This talent had a distinct survival value in that it was thereby easier to hunt at night and not get lost on the way back to camp. Accordingly, the ancestor produced more than his or her share of offspring, and the trait spread rapidly. Much later (95 000 years later), the trait became valuable in allowing humans to read and write.

What is a reasonable response to my "story"? Actually, several retorts are warranted. (a) There is absolutely no hard evidence that language centers in the brain are associated with other non-linguistic abilities. The particular neurological malfunction of dyslexics, for example, is not accompanied by any physical abnormality. (b) We do not know how many genetic modifications were required to rewire the brain so as to allow distinct grammar sections. Multiple genetic changes were likely required, and this greatly complicates any attempts, such as the one above, to correlate the development of two traits, one useful and the other useless. (c) My "story" is vague as to how the gene(s) for star gazing/writing managed to become distributed from the original recipient to each and every human being in the world. (d) By my arbitrarily piggybacking an initially useless trait onto a useful one, I extend the scope of Darwinism to the point of total paralysis. Any seemingly useless trait in Nature could, by a similar argument, be attributed to any obviously useful trait. All predictive value of the theory, such as it is, is thereby lost, and we descend into the realm of metaphysics.

The vowel and verb centers are just two of a multitude of language processing zones widely distributed in the brain and arranged, it is believed, differently from person to person. The presence of distinct verbal centers is also strikingly illustrated by a stroke victim who reportedly could easily name man-made objects such as saws, screwdrivers, and shovels. On the other hand, ducks, camels, foxes, and other animals were indistinguishable to the patient. Apparently, the brain region dealing with natural objects was damaged by the stroke, whereas the region handling man-made objects remained intact. Why Nature should have given us separate zones for natural and man-made objects is not clear. (One wonders where Mickey Mouse, both man-made and animal, is stored in the brain!). Strokes can also knock out a native language but leave a language learned later in life unaffected (and vice versa). Even regular and irregular verbs seem to be processed in different centers of the brain.

If distinct zones for verbs, prepositions, vowels, past tense, man-made objects, etc. are scattered about the brain, as they seem to be, then a mechanism must exist to coordinate the zones in order to formulate a language. No one has any idea how neurons of the disparate zones become activated in a coordinated fashion to generate even the simplest of sentences. We do know that learning a second language increases the density of gray matter in the left inferior parietal cortex, and that the degree of structural reorganization in this region depends upon the attained proficiency as well as the age at which the second language was acquired. But we know next to nothing as to how and when such complex physiology ever evolved. Humans are incredibly smart, and we do not know why.

Part 3: Musical Intelligence

Throughout this book I have been careful to define terminology, but this hardly seems necessary for the word "music". Melody and rhythm are its principal elements, and the ear, vocal chords, and hands are its principal organs of production. The universal popularity of music lies, no doubt, in its ability to arouse pleasure and, on occasion, intense emotional feelings, the nature of which is not understood from a neurological standpoint.

No human talent arises earlier in life than music. It has been claimed that infants as young as two months are able to match the pitch and melodic contour of their mother's songs. Rhythmic structures can be imitated at four months. Creative sound play is evident well before the first words have taken hold. By the age of five or six, most children can manage fairly respectable renditions of the songs to which they have been exposed. Although true musical genius is rare, a reasonably high degree of musical talent is widespread throughout the world's population. Most high schools, for example, have no trouble fielding a football band replete with trombones, clarinets, tubas, and drums. Music, like language, is an inherent component of our intelligence.

The innate musicality of the U.S. populace would be even more evident if our educational system did not regard music as a rather low priority item. In contrast, the Anang people of Nigeria hold music in high esteem. Week-old infants are introduced to music and dance by their mothers. At the age of two, children join cultural groups where they are instructed further in singing, dancing, and the playing of instruments. An Anang child of five can sing hundreds of songs and perform dozens of intricate dance movements. Anthropologists who have studied the Anang, claim never to have encountered a non-musical member.

It is perhaps time to mention the controversial matter of Nature vs. nurture (a topic discussed in more detail in *Chapter 9* dealing with culture). There is no doubt that musical aptitude is genetically linked ("Nature"), and it is either fostered or hindered by the environment ("nurture"). In defense of a natural inherited aptitude, one can cite the Bach family in which great musical art extended for seven generations and included cantors, organists, town musicians, conductors, and composers (eleven of whom achieved eminence in the field). The Couperins are another example of an outstanding multigenerational musical dynasty. Of course, it is difficult here to rigorously separate Nature (i.e. the family genes) from nurture (i.e. the musical tradition in their homes). More persuasive might be the numerous cases (e.g. Arthur Rubinstein) who, in a musically uninteresting home, showed early signs of unusual talent and ultimately rose to lofty heights in the musical world. Such people succeed because they have fortuitously acquired the natural aptitude to become musicians a cut above everybody else. The correct genetic makeup is necessary but not sufficient, however. Only with proper training, hard work, an appealing stage presence, and — yes — luck does their talent become fully manifested in terms of public acclaim. Clearly, both Nature and nurture play important roles here. And, clearly, all the positive nurture in the world would not produce a great musician if the requisite genes were not present.

I will ignore here the debate on Nature vs. nurture because, from an evolutionary standpoint, I tend to interrelate the two. Even if playing the piano expertly required twelve years of instruction by a master teacher ("nurture"), success would in the end be predicated upon a genetic constitution that allowed it. After all, no chimpanzee will ever play the Moonlight Sonata, twelve years of schooling at Julliard notwithstanding. So it makes little difference, at least for the purposes of this book, whether a trait is fully formed at birth or whether a trait must be developed and refined later in life via the assistance of other humans. Either way, it has all been made possible by a remarkable set of genes that evolution has imparted to us. The key question then becomes: How is it that we have become hard-wired to sing opera or play the violin? What is the origin of our extraordinary musical talent?

In reality, musical talent should be subdivided into two distinct types: "Creative" and "reproductive-interpretive". The former entails the writing of new music, while the latter involves the instrumental or vocal expression of music that has already been written by others. The two components of musical intelligence are usually not found side-by-side in the same individual. Few great instrumentalists also produce outstanding compositions, and few great composers are also eminent instrumentalists. Creative musical talent is relatively rare and only occasionally, as in the case of Mozart and Schubert, does it emerge at an early age. Mozart in his eighth year wrote six sonatas for piano, violin, and cello (K. 10–15) as well as a symphony for small orchestra (K. 16). By the age of sixteen, Schubert had written over 100 Lieder in addition to works for the stage, masses, string quartets, and sonatas. But these are exceptions on which I will not dwell because this book focuses on intelligence as distributed more generally throughout the population. In this manner, I protect myself from accusations that I discuss genetic aberrations that crop up only now and then as opposed to musical trends that more generally reflect evolutionary processes that have taken place in the past.

As mentioned above, "reproductive-interpretive" musical intelligence is commonplace and evident even in infants. Musically trained psychologists have attempted to quantify the prevalence and magnitude of musical talent via testing procedures. Sense of rhythm, the ability to grasp and sing a melodic line, musical memory, and analysis of two-note chords have all been included in such testings. Although I am in no position to assess the results, there is one datum arising from the work that is particularly relevant to the present discussion: The number of unmusical persons. In severely unmusical individuals, consonances cannot be differentiated from dissonances; tonality is non-existent; musical compositions cannot be remembered; different melodies of similar rhythm are confused; and there is no recognition of the relationship between music and mood. Studies of school children in Paris have shown that about 10% are unmusical. Among 1000 German students, about 65% were found to be good singers; 30%, mediocre; and 5%, unmusical. Although the exact percentages will depend upon the nature of the test and upon the definitions of "good" and "bad", one thing is clear: The majority of the population is reasonably musical. Thus, in discussing musical intelligence, I am not referring to a trait that appears only occasionally. I am referring to a trait that evolution — for whatever reason — has been inserted into the genetic makeup of almost everyone.

Since we are universally endowed with ears, voice, hands, and allied neurological components or "musical intelligence", we can, with proper training, perform difficult music skillfully if not professionally. But even a moderately skillful music performance demands highly perfected organs of music production. How did such a capacity evolve? How did we happen to acquire the neural connections (both hard-wired before birth and further developed after birth with the aid of education) to make wonderful music? If Darwinism is to be invoked, then one must conclude that human musicality was associated with considerable survival value. Moreover, one must conclude that prehistoric singing and instrument playing entailed, more or less, the same remarkable musical capacity that we treasure so much today. Otherwise, why should modern man now possess such a capacity? These issues, like those of language, are puzzling and merit further attention. I begin with a brief discussion of the human ear.

The human ear is an absolutely astounding piece of biological engineering, but I will spare the reader a description of the anatomical complexities save for the middle ear. The middle ear mechanically transmits via three small interconnected bones called the *ossicles*, vibrations from the eardrum to the so-called oval window. The maximum displacement of the eardrum for the loudest sound that the ear can withstand is about 0.25 millimeters. The minimum displacement of the eardrum at the threshold of audibility is about 2×10^{-9} millimeters. In other words, the ear can detect a sound when the eardrum moves no more than 1/100 the diameter of an atom! Overall, the ear covers a range of about 10^{12} in sound intensity. Consider, for example, several sounds whose loudness is recorded here in a logarithmic unit called the deciBel (dB): Threshold of hearing, 0 dB; library, 30 dB; freeway traffic, 75 dB; and a jet takeoff from 30 meters away, 120 dB. The span from 0 to 120 dB constitutes a trillion-fold audibility-range for sound intensity.

The range of frequencies heard by the human ear varies from about 15 Hz to 15 000 Hz (where a Hertz or Hz equals one vibration per second). Shown below are the frequencies of several piano notes:

Note	Frequency (Hz)	Comment
C	262	Middle C
D	294	Next note upscale
E	329	Next note
A	440	Frequency "standard"
C	523	High C
C	1046	Next higher C

A piano's frequency varies from 28 Hz on the low end all the way up to 4186 Hz on the high end. These values are given here merely to orient the reader toward the Hz frequency unit, which is important in the text below.

The average human ear can hear changes in frequencies of about 3 Hz or greater at 400 Hz. Thus, 395 Hz and 400 Hz, for example, will "sound" different to the average person. A well-trained ear can do even better: 1 Hz at 500 Hz. This means that human beings are capable of hearing about 1/30th of the interval between two adjacent notes like C and D. Clearly, evolution has created in the human ear a capacity to discriminate between two remarkably similar frequencies. Musicians make use of this sensitivity by deliberately modulating the pitch of their notes (a "vibrato"). Violinists, for example, routinely move strings sideways back-and-forth to impose a 5–6 Hz frequency variation upon their notes.

Absolute or perfect pitch is a natural gift that allows a person to identify a note in isolation (i.e. without having heard another note as a guide). Many people with absolute pitch can also sing, for example, an accurate "F" from memory and without reference to another prior note. According to an American study, about 5% of a group of musicians (students, orchestra players, etc.) possess absolute pitch. Single isolated notes from a piano are easiest to recognize; then comes the violin, wind instruments, and finally the human voice. Sung notes offer the most difficulty because identification is hampered by the presence of fewer harmonic overtones (other frequencies mixed in with the main note).

Recent investigations leave no doubt that absolute pitch is a genetically linked trait. In musically gifted children, one finds it as early as the third year. Of course, experience and practice can improve the precision. Absolute pitch is by no means a prerequisite for a musical career. In fact, it can cause confusion when a singer or violinist with absolute pitch is accompanied by a piano tuned even a quarter of a tone too high or too low.

Many musicians without absolute pitch can, nonetheless, recognize and reproduce one particular note. Violinists, for example, acquire a "standard absolute pitch" for the note "A" to which they tune their instrument. Through long practice, there is a gradual and progressive fixation of this single note in the memory. The precision is such that an impression may be given that the musician has a genuine absolute pitch.

Although an absolute pitch is not necessary for a musical career, a relative pitch is absolutely essential. In relative pitch, the relationship between two notes is apprehended through the interval between them rather than from the two individual notes at either end of the interval. An intervallic sense is an inherent gift, common in the population and refined by experience and practice. Chord identification, in which three component notes of a chord are discerned, is another inherited musical talent. The ability to distinguish the component notes of chords has been observed even in children early in their musical training.

Studies of stroke victims show that the majority of musical capacities, including the sensitivity to pitch, are localized in the right hemisphere of the brain. This contrasts with linguistic skills which, as mentioned in the previous section, are found largely on the left side. The situation is more complicated with music than with language in that there is evidence that difficult musical tasks encountered by trained musicians also make use of the left hemisphere. The more a musician is trained, the more likely he or she will draw at least partially from the left side. Certain skills cross hemispheres, some do not. For example, chord analysis seems to be handled always by the right hemisphere no matter how extensive the training. The neurology of musical talent is obscured by the facts that music skills vary a great deal, and that the mode of exposure to music can differ so much, from person to person. Perhaps the most important generalization for our purposes is that damage to the music centers in the brain does not inevitably lead to loss of function in language, mathematics, special skills, etc. Musical neural connections are given their own special corner of the brain.

I return now to the evolutionary basis of the human being's musical ear. How is it that we are blessed with such an incredible organ? At what point in our evolution, and for what survival purpose, did we acquire the ability to distinguish 500 Hz from 504 Hz or (for some of us) the ability to sing a pure "F" note on command? As always, it is possible to construct Darwinian "stories" of explanation: Perhaps we are able to differentiate 500 Hz from 504 Hz because it once helped us detect a prey rustling in the grass. Or perhaps men were once sexually attracted to women with absolute pitch. Or perhaps chord identification helped mothers locate a crying child when the infant crawled outside of camp. Do I believe such "stories"? It is irrelevant whether I or anyone else chooses to believe them. The point is that the "stories" are impossible to prove or disprove. Like snack food, the "stories" satisfy only temporarily; they are not the stuff of science. There is only one truly honest answer to the question "How did we happen to have evolved an ear that allows us to create violin music and grasp its intricacies?" We do not know.

Actually, the human ear, puzzling as it might be to evolutionists, does not present a dilemma equivalent to that of the human voice. With regard to hearing, at least one can point to animal ears that are even more sensitive, if not more "musical", than ours. The human voice is, however, unique in its ability to create sound. Darwinian "stories" attempting to explain the human voice degenerate from implausibility to foolishness. I will finish this section on musical intelligence by delving into how the human vocal apparatus allows us to sing. Naturally, human hearing and singing are interdependent skills that must have required some sort of coordinated evolutionary mechanism. But since nothing is known about how hearing and singing evolved side-by-side, I will treat singing as an isolated component of musical intelligence.

The human voice system consists of the lungs that supply the power (i.e. the air stream); the vocal cords that oscillate to produce a sound wave; and the vocal tract (i.e. the larynx, the pharynx, and the mouth) that serves as a resonating chamber much like the wooden body of a violin.

A sound wave produced directly at the vocal cords is rather simple in form: A strong "fundamental" frequency plus many higher frequencies whose amplitudes decrease uniformly with their distance from the main tone. The situation changes significantly, however, after the sound wave has passed through the vocal tract. The "amplitude vs. frequency" graph is now distorted to include several distinct peaks called "formants". A set of high-strength peaks (the formants) in a typical male voice may appear at 500 Hz, 1500 Hz, 2500 Hz, and 3500 Hz after having passed through the vocal tract. A simplified review of formant theory is necessary to understand operatic singing.

The first point to make is that speaking each vowel generates its own particular set of formants. For example, speaking "*ee*" as in "heep" produces a series of formants, the first two of which lie at about 270 Hz and 2300 Hz. This is another way of saying that "*ee*" gives maximum sound intensities at frequencies of 270 Hz and 2300 Hz. The vowel "*eh*" as in "head" has, in contrast, formants at 500 Hz and 1850 Hz.

There is a second important principle in formant theory: The closer the frequency a desired note is to a formant, the greater the amplitude of the sound that will emerge from the lips. Consider, for example, a singer who wants to sing middle-C (frequency = 262 Hz) and then the C two octaves higher (frequency = 1047 Hz). Assume further that the notes will be sung while voicing a vowel whose first formant lies at 300 Hz. Middle-C will, therefore, nearly coincide with the vowel formant, whereas a C two octaves higher will not be similarly reinforced. The result will be that the higher C has a weaker intensity, and it will have a different quality or "timbre" compared to middle-C. This is a bad situation. The same vowel sung at two different pitches will sound uneven and unpleasant.

Take another example of a problem facing a soprano who, typically, sings in the range of 262–1047 Hz. The C in the middle of her range has a frequency of 523 Hz. If she attempts to sing "*ee*" as in "heed", with its formant at 270 Hz, the C will sound weak because 270 Hz and 523 Hz are far apart. No such problem, however, exists when the soprano sings "*ah*" as in "father" in high F. In this case, the vowel has its first formant at 700 Hz, and the F note has a nearby frequency of 698 Hz.

"ee" in C → "ah" in F
(weak) (strong)

But let us return to the soprano who wants to sing "*ee*" with the note C. She could, of course, modulate the air pressure, etc. in order to compensate for the inequities in intensity and timbre, but this is difficult and exhausting to do in practice. It is much easier to move the first formant upward to match the frequency of the fundamental. Thus, when a soprano sings "*ee*" in C, she actually moves the vowel formant at 270 Hz closer to the C frequency at 523 Hz. How is this done? The usual way is to open the mouth wider or, alternatively, to shorten the vocal tract by drawing back the corners of the mouth. This draws the "*ee*" formant closer to the 523 Hz note, and the note intensity increases.

An "*ee*" that is sung sounds differently from a spoken "*ee*" owing to their different formant compositions, but this "funny pronunciation" cannot be helped. A distorted pronunciation is the price one pays for a uniform volume and timbre among various notes.

A second problem faces the opera singer, namely the need to compete with the accompanying orchestra for audibility. In order not to be drowned out by the orchestra, opera singers learn how to "lower the larynx". A lowered larynx, being acoustically mismatched with the rest of the vocal tract, generates a 2500–3000 Hz frequency of its own. This new 2500–3000 Hz peak, called the "singer's formant", is inserted between the third and fourth formant normally present. Now since an orchestra is loudest at 450 Hz, the new 2500–3000 Hz formant helps the singer rise above the accompaniment. Lowering the larynx also darkens the vowels, another reason why in opera sung vowels have an "unusual" pronunciation. With pop singers, who make use of electrical amplification in order to be heard above the band, ordinary spoken vowels are satisfactory.

No one knows how, evolutionarily speaking, the remarkable ability to sing opera (i.e. to shift and create formants) appeared on the list of human talents. The same can be said of other musical skills such as violin playing. Experts believe that with proper training most people could sing operatically. There is nothing freakish about the anatomy and physiology of accomplished opera singers; one cannot predict, for example, who will be a successful opera singer by means of a CAT-scan. This does not mean we are all cut out for operatic careers any more than we are all cut out to be professional tennis players. Other factors, such as a refined sense of "musicality", enter the picture. But Nature has certainly endowed all of us with the basic equipment, and we do not know why we have been blessed in this manner. Ordinary speech has never required operatic tones. Nor did ordinary music (as might have been performed by Freddy Flintstone and his Neolithics, if I might be allowed a bit of silliness) require operatic tones. Perhaps a vocal system permitting the manipulation of formants helped prehistoric man to shout great distances during communal hunts. Those who did this best reproduced themselves preferentially. And perhaps basket-weaving ultimately endowed humans with the manual dexterity and intellect of a piano player executing a work by Rachmaninoff. And perhaps listening to snakes in the grass led to our present auditory ability to hit frequencies precisely on a violin. The point here is that (as was already evident in our journey through the linguistic component of human intelligence) we are being continually forced to pile story upon story, uncertainty upon uncertainty, mystery upon mystery. It becomes unsettling even to an ardent Darwinist. One yearns for a rigorous explanation of those human capabilities that enrich our lives in modern times but have no proven reason for existence in the prehistoric past.

The reader may legitimately ask of me: "Are you merely trying to prove that we are deficient in our understanding of evolution? If so, then you are proving what every thoughtful person knows or at least suspects." My response to this would be: Yes, I am attempting to prove, via consideration of human intelligence, that we are substantially ignorant in our understanding of evolution. By so doing, I hope to counter in some small measure the contention that natural selection is, for all practical purposes, a closed book, a solved problem. Although only a few texts dare state such a conclusion explicitly, many infer a "case closed" position by suppressing the uncertainties that plague the field. To some extent, unswerving dogmatism has been motivated by a justifiable pride in the wonderful scientific advances of the 20th century. Emotional confrontations with creationists have also hardened positions on both sides and made it more difficult to confess to uncertainty. But it is just as misleading to claim that we understand more than we do as it is to claim that we understand less than we do. Even if my sole objective were, therefore, to proclaim a certain level of ignorance, my writings would not be pointless.

I do not mean to assert that natural selection is incorrect (and under no circumstances should my treatise be misused to support an anti-Darwin agenda in which evolution itself is questioned). But when an organism possesses a complex, multigene, energy-draining trait that far exceeds any plausible prehistoric need, we must search beyond neo-Darwinism for an explanation. The varied abilities housed within our "thin bone vault" constitute such a trait.

Part 4: Mathematical Intelligence

Professor A. C. Aitken of Edinburgh University was once asked to divide 4 by 47 in his head. After 4 seconds, he responded with the following set of numbers (taking less than 1 second per number):

0.08510638297872340425531914

He stopped, paused a minute, and then continued the sequence:

191489 … (5 sec pause) … 361702127659574468 …

Aitken then said: "Now that's the repeating point. It starts again with 0.0851, etc."

How did such a brain evolve? What Paleolithic need ever demanded such a facility with numbers? Admittedly, Aitken's arithmetic talents appear but rarely in the population. Yet the fact that they appear at all suggests that the rest of us, with more normal arithmetic capabilities, may be on the edge — evolutionarily speaking — of duplicating Aitken's performance. Whether this is true or not, human beings are clearly the owners of an unfathomably refined mathematical intelligence, and one need not recount a bizarre numerical ability to prove it. All one must do is examine the student body of a typical university.

In a given semester at my university, about 19% of the undergraduate students take beginning calculus. If this percentage is similar to that in other universities, then roughly 2 600 000 students nationwide enroll in calculus. A goodly proportion of these, perhaps 1 500 000 or more, master the subject reasonably proficiently. This number does not include the multitude who are perfectly capable of learning calculus but choose, instead, to study other subjects. Since the United States produces millions and millions of people who can understand and manipulate the tenets of calculus, the ability to do so, while not exactly commonplace, is certainly unexceptional. Evolution has produced a brain that allows a large section of the population to handle abstract mathematical concepts.

I selected calculus as an example for a reason other than its prevalence. I selected it because unlike arithmetic, calculus deals with non-numerical symbols. Even "simplest" of expressions in calculus (e.g. the definitions of a derivative and integral found in all elementary calculus texts) demonstrate that calculus is a mental discipline that computes via symbols other than numbers. Although our prehistoric ancestors may have had certain arithmetic abilities (counting, adding, subtracting, etc.), no one (to my knowledge) has been rash enough to suggest that our ancestors ever solved mathematical problems symbolically. So college students using calculus (or, for that matter, high school students using algebra) are engaged in an abstruse mental exercise that was irrelevant to prehistoric life and thus to human evolution in general.

Consider one specific example of symbolic mathematics. An ex-postdoctorate of mine from Holland, Eric van der Linden, developed at Emory an extended theory of lamellar hydrodynamics in which he derived the relaxation time for undulation in a three-component membrane system. (This is not supposed to mean anything to non-physical scientists, but those interested can consult *Langmuir*, 9, 692, 1993). The point here is that 10 000 years ago Eric's ancestors were hunter-gatherers. While in no way wanting to demean the intellectual needs for hunting-gathering in humans (or in any other hunting-gathering animal for that matter), it would be absurd to suggest that Eric's ancestors ever solved mathematical problems symbolically. From where, then, did Eric's obvious mathematical talents evolve? From where did abstruse mental skills possessed by all those students of calculus and differential equations derive?

$$\tau_0 = \left(\frac{1}{2}\right)\{4\eta_0 q\}\left\{\left(\gamma + \left(\frac{K}{2}\right)C_0^2\right)q^2 + Kq^4\right\}^{-1}$$
$$\times\left\{\left[\frac{\cosh qd_0 - 1}{\sinh qd_0 + qd_0}\right] + \frac{(\eta_0/\eta_0)[\cosh qd_w - 1]}{\sinh qd_w + qd_w}\right\}$$

A brain that can think in abstract mathematical relationships is an "organ of perfection" like the eye. But — and this is an important "but" — there is a huge evolutionary difference between the eye and the brain. In the case of the eye, the survival value of its function, sight, is obvious. Thus, the evolution of the eye, although not understood in detail, can at least be postulated as the product of natural selection. This is not possible with the brain's ability to think in abstract mathematical terms. Esoteric mathematical gymnastics were simply not part and parcel of our survival needs when we evolved as primitive hunter-gatherers. I repeat an assertion made in the sections on written language and music: A complex trait of no proven utility violates the essence of natural selection.

We now arrive at a difficult question: How much of our ability to think in terms of complex symbolic expressions, such as Eric's equation, can be classified as "cultural"? There is no doubt that our current grasp of calculus stems from two factors: (a) the original development of the field (starting with Leibnitz and Newton three centuries ago) and (b) the teaching of the discovered principles and concepts to succeeding generations. Both of these factors, discovery and teaching, are at least partially "cultural". My contention, however, is that culture is necessary but not sufficient and, in this regard, mathematics is no different from language and music. Calculus could not have been discovered and taught to others were there no brilliant minds capable of extremely creative thinking. But millions of others could not have learned the concepts, once uncovered, unless their brains could absorb complex thoughts in symbols largely foreign to any linguistic association. We are smart not only because our culture provides us good books and teachers (although, as a teacher, I like to think this helps!). We are smart because our brain allows us to be smart, pure and simple.

To summarize the argument thus far: (a) Any trait (especially any complex, energy-consuming trait) must be in some manner needed and useful or else it would be unaccountable by natural selection. (b) Prehistoric man had no need for abstract mathematical thinking as manifested, for example, by differential and integral calculus. (c) Therefore, abstract mathematical thinking has origins independent of natural selection. This syllogism rests on the assumption that prehistoric man did indeed have no use for solving difficult quantitative problems in symbolic representations (or for its equivalence in some aspect of life). The validity of this assumption can best be assessed from the mathematical ability of modern hunter-gatherer societies. If modern hunter-gatherers do not presently make use of abstract mathematical thinking, it hardly seems possible that prehistoric man did otherwise in the distant past. Were this false, one would have to presume that prehistoric man was intellectually more sophisticated than modern primitive man. It was this line of reasoning that led me to inquire as to exactly what type of mathematical operations fulfill the needs of modern primitive people. Here are a few sample facts I gathered from the anthropological literature on modern primitive people before Western man imposed its influence:

a) Tasmanian Natives: These nomadic hunters, being ignorant of agriculture, sought food by collecting, fishing, and hunting. They could count to no higher than 5.
b) The Crows of the Western Plains: It was reported that "they do not usually count higher than 1000, as they say that honest people have no use for larger numerals."
c) Ainus of Japan: Although their spoken language embraces about 14 000 words, they had no form of writing or even picture-writing. Their numerical system was "vigesumal", i.e. based on 20 rather than 10. The highest number was 800. Units of measurement were exceedingly simple: Distance was reckoned in steps; lengths in spans; quantities in handfuls, and weight and area not at all.
d) Aranda of Central Australia: Practicing no agriculture, they survived by foraging and hunting. Time was reckoned in "sleeps" and "moons", and an erudite native could count to as high as 5.
e) Dahomeans of West Africa: These people lived in the hot, damp tropical climate of a luxuriant jungle. They used a quinary-vigesimal system of numeration based on 5 and 20, but for large numbers they preferred terminology derived from strings of cowries (shells) with 40 cowries per string. A set of 50 strings

was called a "hoto", and this same word was used for the number 2000.

f) Pirahã of the Maici River in Brazil: They currently use a system of counting called "one-two-many". The word for "one" means "roughly one" (equivalent to our "one or two"). The word for "two" means "a slightly larger amount than one" (equivalent to our "a few"). And the word for "many" means "a much larger amount".
g) Semang of the Malay Peninsula: This race of pygmies with chocolate brown skin possessed no form of writing and never counted beyond 3.
h) Thongas of South Africa: Their system of numeration was distinctly decimal with words for only seven numbers: 1, 2, 3, 4, 5, 10, and 100. All numbers were expressed with these seven words (e.g. 987 = 5–100's, 4–100's, 5–10's, 3–10's, 5, and 2). Natives generally used their fingers in an unusual sequence when counting:

Number	Hand (s)	Finger (s)
1	left	5th
2	left	5th and 3rd
3	left	5th, 3rd, and 2nd
4	left	5th, 3rd, 2nd, and 4th
5	left	all five fingers
6	both	left hand and right thumb
7	both	left hand, thumb, and 2nd
10	both	clap two hands

This was about as high a development of numbers reached by the Thongas who did not much trouble themselves with larger numbers. They were quick to declare that a very large number was "one which passes the capacity of the reckoners" and to use the word "tjandjabahlayi" meaning "innumerable". Henry A. Junod, who studied the tribe intensively, wrote in his 1962 book entitled "*The Life of a South African Tribe*" the following: "So, on the whole, opportunities of using the arithmetic facility are very rare in primitive native life, and we ought not to be astonished that this faculty has remained undeveloped. To pretend, however, that it is altogether wanting would be erroneous. I can give many proofs that it exists, and sometimes manifests itself in interesting ways." Among other manifestations of mathematical skills, Junod mentions the successful teaching of algebra in missionary schools.

Recent work probing the inherent mathematical ability of hunter-gatherers has shown that, in the absence of specific words for numbers, number-sense grows fuzzy beyond the numbers 3 or 4. Research by a team of French cognitive scientists on the Mundurukú (a group of Indians living in a remote section of the Amazon) suggests that core geometric concepts are part of the basic human mental apparatus. Thus, the Indians correctly read simple maps and could select a geometric figure that "did not belong" from a set of six figures roughly 70% of the time. One cannot, of course, easily differentiate this so-called "shared core of geometric knowledge" from a general reasoning ability.

What can one conclude from such observations (and many others like it that I have omitted for the sake of brevity)? Several points seem clear: (a) The concept of number is universal. (b) The use of numbers among primitive people is rather elementary. For example, counting often does not exceed the number of fingers and toes, and basic arithmetical operations such as multiplication and division are most often non-existent. (c) Finally, and most pertinent to the general discussion of evolution, it is obvious that primitive people (clever folks living in all sorts of challenging environmental situations such as the jungle, desert, tundra, prairie, etc.) thrive as hunter-gatherers without requiring abstract mathematical concepts or, for that matter, even rather elementary arithmetical operations such as long division.

Classical evolutionists of the late 19th century totally misinterpreted the apparent "limited" intellectual abilities of traditional peoples. It was taken for granted that these peoples were "living ancestors". They were considered (as were our hunter-gatherer ancestors) lowly, ancient, and primeval in intellect. They were viewed as children because childhood is also an early stage of human life. They were often called "prelogical" in contrast to ourselves of the Western culture who are "logical". Traditional peoples were ostensibly of lesser intelligence, capable of concrete thought but not capable of analytic or abstract thought.

The above portrayal of primitive peoples is grossly false. Perhaps the most telling evidence for the unacceptability of the early evolutionist picture relates to the remarks that Henry Junod made about the Thongas of South America. As stated above, Thonga children, when taken out of their culture and placed in Western-style schools, can learn algebra despite their virtually mathematics-free heritage. The mathematical trainability of displaced traditional peoples is a generally observed phenomenon. Thus, Thongas do not do algebra because they do not need algebra. They live fine without it. Their mental wiring can be organized, however, to accommodate abstract mathematical terms should the occasion (such as a missionary school) ever arise.

Marcia Ascher in her wonderful book entitled "*Ethnomathematics*" describes activities of traditional people that have a mathematical component to them. These include games of strategy and chance, the tracing of complex designs in the sand, and the organization and modeling of space (as involved in navigation and in the artistic creation of intricate "strip decorations"). One specific example of a game once played by the Cayuga Indians of northeast United States will be recounted here:

The game consisted of a wooden bowl plus six smoothed and flattened peach stones blackened by burning on one side only. The six stones in the bowl were tossed in the air. Similar to the game of dice, the stones could land with six black sides (B's) facing up, or six neutral sides (N's) facing up, or any combination in between. Scoring was according to the table below. Thus, 6 B's or 6 N's gave 5 points. Five faces of the same color gave a player 1 point. Any other combination gave no points at all. If a player scored any point, he or she earned another toss. Some prearranged total number of points, ranging from 40 to 100, constituted victory.

Outcome	Number of Points	Statistical Factor
6 B's	5	5
5 B's and 1 N	1	5/6
4 B's and 2 N's	0	1/3
3 B's and 3 N's	0	1/4
2 B's and 4 N's	0	1/3
1 B and 5 N's	1	5/6
6 N's	5	5

As seen in the table, the number of points that the Cayugas awarded to each particular combination of stones closely parallels the statistical improbability of the combination. Many additional examples cited by Ascher likewise point, at some level, to an intuitive grasp of mathematical concepts. Almost invariably, however, these examples relate to games and artwork rather than to evolution-promoting activities essential to reproduction and survival.

The inescapable conclusion is that primitive people possess an innate mathematical intelligence that, owing to lack of need, never becomes fully expressed. Indeed, if we have learned one critical fact about modern primitives it is that when transplanted at a young age to an advanced society, they can learn math and other subjects as easily as anyone else. This was presumably true for prehistoric man as well (although it is unclear how far back in time the term "prehistoric" can be pushed). But if mathematical intelligence were unutilized, or at least underutilized, in the era of hunting-gathering (an era that comprised all but the final 10 000 years of our evolution), then by the laws of natural selection we should not possess it. Thus, mathematical intelligence joins written language and musical ability as unsolved evolutionary mysteries.

One may be tempted to advance a counter-argument, namely that mathematical talent is nothing special; it is simply a modern expression of skills required by hunter-gatherers to survive.

Responding to this assertion is not easy. After all, prehistoric man had to be clever to survive. He had to outwit game animals; to distinguish edible from inedible plants; to make tools; to communicate. In no instance, however, have I ever seen a proof that the cleverness attached to these activities translates smoothly into abstract mathematical thinking. This is not to say that there is absolutely no overlap between daily activities of hunter-gatherers and mathematics. For example, deducing the proximity of a game animal from its footprints demands logic as does, of course, solving a calculus problem. But the similarity stops there. A footprint is tangible; one can touch it; smell it; gaze at it. Most mathematical expressions lack any connection with an ordinary person's palpable reality and mundane existence. No, mathematical intelligence is more than a simple application of common, everyday thinking. It is unique and marvelous gift of unknown origin.

The preceding paragraph brings up an important proven concept in brain performance: *Plasticity*. As has been pointed out repeatedly, the brain is plastic in that it can accommodate changes if demands are placed upon it. A common example is that of a blind person whose sense of hearing becomes far more acute owing to rewiring of brain connections. Thus, an argument can be advanced that our mathematical ability reflects brain capacity originally developed for other purposes. We do not know what these purposes are. Language does not seem to be a likely surrogate because the connection between language and mathematics (e.g. Eric's equation) is tenuous. Nor is hunting, of the type carried out skillfully by a variety of other animals, an appealing forerunner of mathematics. Tool-making no doubt provides evidence for prehistoric intelligence, creativity, and culture, but tool-making is also a manual activity with survival value, in sharp contrast to high-level and abstract mathematical thinking. Most bothersome of all: To propose an evolutionary connection between two diverse traits, in an ad hoc manner devoid of evidence, is to allow evolutionary explanation to deteriorate into utter meaningless. Is an understanding of evolution elevated by claiming that our ability to understand astrophysics originates from an opposed thumb? Or that our ability to play the violin, and to conduct brain surgery, originates from the making of pottery? It is far better to be candid and admit our ignorance in the subject.

Despite my reluctance to equate unrelated mental abilities, the fact remains that plasticity is an important attribute of the brain. Yet, in a sense, plasticity and intelligence are the same thing. Rather than clarifying the intelligence problem, "plasticity" merely restates it. Thus, we want to know why humans are so smart. This desire can also be expressed as wanting to know how we happen to have a brain with such an enormous capacity and plasticity that it serves us well even when it comes to esoteric mathematical challenges with no evolutionary history.

Swiss psychologist Jean Piaget advanced a theory on how mathematical intelligence unfolds in human beings which I now recount.

Mathematical intelligence derives most fundamentally from the observation and manipulation of physical objects beginning with those in the nursery. The infant explores these objects (i.e. gazes upon them, touches them, chews them), but as soon as they disappear from sight, they no longer occupy the infant's consciousness. An object in view is only a transitory experience.

At about 18 months of age, the child begins to appreciate an object's existence even after it has been removed from sight. This, the awareness of "object permanence", is a milestone in our mental development because it allows a child to think of objects in their absence. Out of sight is not necessarily out of mind.

"Categorization" is another milestone, although in the first few years of life the categories are qualitative rather than quantitative. A child can recognize which of two piles of blocks is the taller, but he has no concept of the number of blocks in each pile. He will select, likely as not, three candies spread over a large area in preference to five candies confined to a small area (space being the critical determinant at this stage of development). Although a child may be able to count out loud, this is more by way of linguistic intelligence than a fundamental understanding of the numbers' significance.

At the age of four or five, a child begins to coordinate numbers with a group of objects. He touches the first object and associates it with "one", a second touched object with "two", etc. Eventually, and this is a pivotal step in the development of mathematical thinking, the final number in the count is taken as the total number of objects. This skill is further refined until the child can count two groups of objects and thereby determine which group possesses the greater number. The concept of comparing numbers and "quantity" becomes ingrained. Note that culture plays a distinct role in the development of a child's innate mathematical tendencies. A child deprived of toys and other interesting objects will not progress as quickly. The neural pathways are there at birth waiting to be developed, but the connections that are actually made depend upon the environment.

Things develop more or less as one would expect. For example, a child comes to realize that counting a row of blocks from left to right gives the same number as counting from right to left. Elements of addition and subtraction enter into the child's play. For example, the child notices that adding two blocks to a pile of four creates a new pile identical to a neighboring pile of six. Removing a block from a pile of six achieves equivalence with adding a block to a pile of four. The beginnings of arithmetic are in the making (often assisted by parents or older siblings).

Eventually, the arithmetic operations become "internalized", i.e. a child can perform them in his mind in the absence of any tangible objects. For example, he can create a mental picture of adding three blocks to another three blocks to form a pile of six blocks. At this early stage of mathematical thinking, mental images still entail discrete objects as opposed to abstract numbers.

Piaget's final stage of cognitive growth is reached in adolescence. The child begins to operate not only on objects but on symbols representing objects. Simultaneously, elements of logic take hold. For example, the child will comprehend a statement to the effect that "If $X = 8$, then $X + 2 = 10$". Drawing logical conclusions by manipulating symbols is the essence of higher mathematics.

Although one might debate the discrete steps in operational thinking that Piaget suggests, the above is a more-or-less accurate account of what happens to a Western middle-class child. Having an idea of what happens does not, however, reveal anything about why it happens. Why is it that a teenager can understand the statement, "If X = 8, then X + 2 = 10" (not to mention far more complicated symbolic expressions)? We do not know the answer to this question. We do know, from experiments such as those of Karen Wynn described below, that even infants possess a capacity to perform simple arithmetic operations.

The experiments on 5-month old infants used a "looking time" procedure that has become standard in studies of infant cognition. It is based on the fact that infants look longer at unexpected events than at expected ones. If, for example, an infant stares at two suddenly-appearing dolls for 20 consecutive seconds before looking away, compared to only 10 seconds for a control, then this indicates that the infants did not expect to see the two dolls. If surprise is elicited when two dolls appear as a result of an illogical arithmetical operation, but not a valid one, then the infant can obviously tell the difference.

The following procedure was carried out: (a) An object was placed in a case in full view of an infant. (b) A screen was lowered to block the view of the object from the infant. (c) A second identical object was then delivered behind the screen. The infant was able to observe the delivery of the object as well as the departure of the empty hand. (d) The screen was suddenly removed to reveal either two objects (a "possible outcome") or a single object (an "impossible outcome"). Surprise by the infant was evident only with the "impossible outcome", indicating an ability to understand that 1 + 1 = 2. A related experiment revealed an ability to distinguish 2 – 1 = 1 (no surprise) from 2 – 1 = 2 (surprise). It was concluded that "humans innately possess the capacity to perform simple arithmetic calculations, which may provide the foundations for the development of further arithmetical knowledge."

Marcia Ascher wrote: "Mathematical ideas involve number, logic, or spatial configuration and, in particular, the combination or organization of these into systems or structures." Many human beings, namely those in the upper range of the population's mathematical intelligence, have the ability to manipulate systems and structures of absolutely astounding complexity. From where does this talent come? We know that infants have an innate proclivity toward numerical operations. We know that modern hunter-gatherers can count and, in certain cases, do simple arithmetic. In no case, however, have mathematical ideas and operations of a truly sophisticated sort ever been found in modern traditional peoples or found necessary for survival. In short, our amazing mathematical talent vastly exceeds any conceivable evolutionary need. We have no "evolutionary right" to be such good mathematicians; or musicians; or writers; or architects; or computer programmers; or auto mechanics; or philosophers.

Lest the reader still not be impressed with the human brain, allow me to cite anecdotal data taken from the Guinness "*Book of World Records*":

a) Barbara Moore (1988) performed on the piano 1852 songs from memory.
b) Bhandanta Vicitsara (1974) recited 16 000 pages of Buddhist canonical texts in Rangoon.
c) Gou Yan-ling memorized more than 15 000 telephone numbers in Harbin, China.
d) William Klein (1981) extracted the 13th root from a 100-digit number in 1 minute and 29 seconds.
e) Shakuntala Devi (1980) multiplied 7 686 369 774 870 by 2 465 099 745 779 (numbers selected at random by a computer). His answer was 18 947 668 177 995 426 462 773 730.

I have pointedly avoided much discussion of "genius" in favor of more commonplace forms of intelligence. In this manner I avoid vulnerability to arguments that I am focusing on "freaks" of little evolutionary import. It is interesting, however, that a certain minority of psychologists believe that with enough effort almost anyone can attain prodigy-level performance in a desired field.

Anders Ericsson, a member of this "practice makes perfect" school, is of the opinion that a highly accessible long-term memory is the essential ingredient of genius. The following case of Rüdiger Gamm, as examined by neuroscientist Nathalie Tzourio-Mazoyer, is consistent with this view:

Gamm can mentally calculate the fifth root of a ten-digit number within seconds or mentally divide two numbers to 60 decimal points. Positron-emission tomography reveals that Gamm uses more of his brain than normal — in particular, three areas that have been previously linked to long-term memory. Presumably, this extra long-term memory space helps him avoid "losing his place" as he grinds out his mental arithmetic. Gamm was not always so skillful. At the age of 20 he began practicing four hours a day until, at the age of 26, he managed these amazing mental feats. If Ericsson is correct, and ordinary people, with sufficient dedication, can elevate themselves to prodigy-level heights, then humans' mental capabilities may encompass a "hidden but exploitable" component and actually be even more amazing than I have thus far inferred.

I have paid no attention to an extensively discussed treasure within the "thin bone vault", namely our consciousness. The exact definition of consciousness depends upon whom you ask, but a "thought experiment", devised by philosopher Frank Jackson, nicely illustrates one aspect of consciousness. Pretend that a certain Dr. Mary Smith, a neuroscientist and the world's leading authority on color vision, has lived her entire life in a black-and-white room. Although she has never seen the colors of the rainbow, she knows everything there is about the physics of visible light, the brain response to the various colors, neurological integration of visual information, etc. Despite this expertise, Mary does not know, and cannot know, what it is like to experience red color. Her knowledge of biophysics and the physiology of the brain does not yield the knowledge obtainable solely from "conscious" experience. Consciousness is another gift of the brain whose origin is cloaked in mystery, but it lies outside the scope of this book.

In summary, I have discussed three of the many components of intelligence: linguistic, musical, and mathematical. When mention was made of, for example, a "music center in the brain", one should not mistakenly picture a section of the brain that, like the eye, has a specific size and location. Instead, a "music center" is a neural network whose size likely depends upon the particular cultural input experienced after birth. Its precise location may vary from person to person. Our dilemma, of course, is to explain why the human brain has, over the ages, made available such intellectual centers in the absence of the cultural input that was needed to stimulate the development of the centers. Dismissing our intellectual centers as having evolved by natural selection because they were genetically connected to some unrelated life activities during our hunting-gathering days is facile, specious, and unproven. We have a huge plastic brain, one that is very expensive to maintain, and the theory of natural selection does little to understand its origins.

Chapter 7

A BRIEF HISTORY OF THE MIND

Having established the mental capabilities of modern man, I can now inquire into intelligence levels at various previous stages of human evolution. As one might imagine, hard information on the subject is sparse. Paleoanthropologists have often referred to brain size as an indication of intelligence. Ostensibly, the bigger the brain per body weight, the smarter the organism. (It is necessary to correct for body weight because otherwise we would be forced to admit that whales and elephants, which have much larger brains than ourselves, are more intelligent). Yet even with this correction, conclusions are dangerous: young chicks and adult mice have brains that are 4% of body weight...twice the brain-to-body weight ratio of humans. With regard to prehistoric man, *Homo habilis*, who lived about 3 million years ago, had a brain size of just under 700 cubic centimeters (cc) compared to an average modern human brain of 1400 cc. Chimpanzees have a 400 cc brain about twice as large as most mammals of their size. Again, things do not fall in place as neatly as one would like. For example, Neanderthals of 100 000 years ago had, on the average, a brain size slightly larger than that of modern man. Human female brains are on average about 15% smaller than those of males with no impairment of intelligence. And Anatole France had a brain size of less than 1000 cc, but this did not stop him from winning the Nobel Prize in literature.

It seems clear that we are born with far more brain capacity than we need (a point that has been made repeatedly). This is nowhere more obvious than in reports such as that of a boy who, owing to atrophy of his right cerebral hemisphere, began suffering from debilitating seizures. At the age of 19, he submitted to surgical removal of the right side of his brain, an operation that actually raised his IQ by 14 points to 142. Apparently, the seizures, which had been alleviated by the operation, had hampered his intelligence although the surgery itself did not do so. Fifteen years later the individual had obtained a university diploma and held a responsible administrative position. Other half-brained persons have also been successful in the business and academic worlds. We are super-endowed with intelligence.

If the brain size per se is not the sole anatomical determinant of intelligence, what is? Clearly, human evolution is associated with major restructuring the brain, size being only one of many parameters. For example, Neanderthal skulls have a lower and flatter forehead, thus positioning their brains toward the back. Modern humans, in contrast, have high, bulging foreheads that house the prefrontal lobes of our brain. Space is available for a super-developed neocortex that is the seat of our intelligence and awareness including thinking, problem solving, reading, and appreciation of beauty and manual skills. Although the human brain is more than three times the size of a chimpanzee brain, we have only 25% more cerebral cortex cells. Human brain cells, however, are larger, more complex, and spaced further apart than in the chimpanzee; the human cells have evolved with thousands of interconnection to each other. All in all, it is the uniquely human neural organization to which we can ascribe our intelligence. We may not be able to precisely define physiologically or biochemically how human mental prowess differs from that of our non-human relatives, but few scientists doubt that evolution has restructured our brains and, as a result, humans have become something more than simply "smart apes". Exactly when this momentous restructuring occurred, and over what time period, no one really knows, but evolutionarily speaking the process began relatively recently, perhaps within the past two millions years, with major advances as recent as 40 000 years ago.

At one time it was fashionable to examine brains of deceased geniuses in an attempt to determine the source of their intelligence. Thus, the Wistar Institute in Philadelphia once had a collection of over 200 human brains including that of poet Walt Whitman. Albert Einstein's brain ended up in the apartment of pathologist Thomas Harvey in Lawrence, Kansas. Since such work showed rather normal human brains (although Einstein's brain had more support cells, called glia, per neuron than average), the approach has been largely abandoned. Incidentally, Walt Whitman's brain, it was reported by a Wistar spokeswoman, was dropped on the floor by a lab technician and discarded.

Tool development has been used to assess human intelligence but, as with brain size, tools have proved less informative than one might have hoped. To illustrate the limitations of tools as a criterion of intelligence, I will begin by comparing modern man with the ancient Greeks who, between 500 and 200 BC, produced such illustrious people such as Euripedes, Socrates, Plato, Aristotle, Euclid, Archimedes, Aristophanes, and Homer. Can there be any doubt that the Greeks were every bit as intelligent as ourselves? True, the Greeks lacked modern "tools" such as the laser, jet engine, and nuclear power plant. But their intellectual accomplishments were as remarkable in relation to when they lived as today's scientific and technological advances. I have not the slightest hesitation in claiming that the ancient Greeks could have readily assimilated the abstract notions of higher mathematics and physics had these been a component of their culture. Indeed, the principles of sequential logic and the manipulation of symbols (e.g. "...and it follows from A that B is true...") originated with the philosophers of the Golden Age in Greece. This occurred even though our brain was not specifically designed to handle linear chains of symbolic reasoning. We may not know when our super-intelligence first appeared, but we do know, at least, that it must have predated the ancient Greeks, and it may well date back to prehistoric times (100 000 years or more ago).

Now let us proceed further back in time (in a greatly simplified history) to the Neanderthals, a people who inhabited Europe from about 230 000 years ago until 30 000 years ago when they vanished. Traditionally, Neanderthals, with their massive bones and thick brow-ridges, were viewed as hulking sub-humans of low intelligence who even lacked the anatomical and intellectual ability to speak. Their artistic and tool-making skills also seemed primitive. For example, Neanderthals made no spears and relied on a "flake technology" (which changed little during their entire tenure on earth) to fashion flint into crude knives, scrappers, and points. Current paleoanthropologists are now somewhat kinder to Neanderthals, admitting as they do that Neanderthals might have spoken a rudimentary language. There are even those, such as Neanderthal specialist Fred Smith, who call the Neanderthal a "highly resourceful and intelligent creature".

In the 1980s, scientists unearthed the remains of an anatomically modern man who had lived about 90 000 years ago. Mitochondrial DNA evidence suggests that early modern humans (*Homo sapiens* also referred to as Cro-Magnons) first entered Europe about 60 000–85 000 years ago from central Africa where they had been living for some 200 000 years (the so-called "Out of Africa" hypothesis). A second wave (people who had originally also left Africa) entered Europe from central Asia about 35 000–40 000 years ago. Coinciding with this latter migration, there appeared sophisticated tools, more efficient hearths, better shelters, and (according to the anatomy of the vocal apparatus and enlargement of certain areas of the brain) a full capability for language. Their advanced tool technology can be exemplified by skillfully made blades and projectile weapons which were long and thin and had ten times more cutting edge per lump of flint than those of the Neanderthal. Archaeologists have also found in Cro-Magnon campsites various body ornaments such as beads and pendants made from soft stone, shell, and ivory. Thus, after some 2.5 million years in which the archaeological record reveals relatively little innovation, there exploded onto the European scene, about 35 000 years ago, new kinds of tools, body ornamentation, and visual images.

Only after this remarkable cultural explosion 35 000–40 000 years ago did the Cro-Magnon, according to this scenario, finally achieve an advantage over their Neanderthal neighbors. The spurt in intelligence has been attributed to "new neurological connections", but this is a rather information-sparse explanation for what was one of the most momentous and progressive events in the history of mankind.

The Neanderthals disappeared in less than 10 000 years after the arrival of culturally advantaged Cro-Magons. Possible reasons for the demise of Neanderthals include: (a) annihilation by the Cro-Magnons; (b) introduction of new and deadly diseases by the Cro-Magnons; (c) a slower die-off due to an inability of Neanderthals to compete; and (d) absorption through interbreeding. The last possibility seems unlikely because mitochondrial DNA tests on Neanderthal tissue do not indicate much genetic mixing with modern humans.

Unfortunately, the wealth of archaeological information on Neanderthal tools has failed to produce definitive conclusions about Neanderthal intelligence. Were it otherwise, experts would be agreeing on the subject. The problem lies, of course, in the difficulty of differentiating intelligence from culture (the well-known "nature vs. nurture" dilemma). One can no more conclude that Neanderthals were dumb brutes because their tools were crude than one can conclude that the Greeks were mentally inferior because they lacked a laser or jet engine. Neanderthals might have produced crude tools for no other reason than advanced tool-making methods had never been discovered and disseminated.

It may appear that I am guilty here of an inconsistency. On the one hand, I claim that tools are ambiguous with regard to Neanderthal intelligence. On the other hand, I have also claimed, on the basis of modern accomplishments, that humans of today are unambiguously and inexplicably intelligent. Yet, in actual fact, the two claims are not at odds with each other. The Neanderthal tool information is "negative" in the sense that we know only that Neanderthals did not produce sophisticated tools, and (as just pointed out) this could have more than one explanation. In contrast, current human accomplishments are "positive"; today's human beings display observable mental feats that would be impossible without an extraordinary intelligence.

Apart from being symptomatic of a culture's sophistication, tools have, some believe, actually played an important role in the evolution process. The argument goes something like this: Early humans, after developing crude tools, became more intelligent in order to make better use of those tools. As humans became more intelligent, they made even better tools that required them to become even more intelligent, and so on. Although the story is a bit simplistic, there is a ring of truth here in that the evolution of intelligence is, no doubt, associated with the human hand. The hand, along with the brain, and the communication apparatus, are the three key elements of human progress.

A history of the hand, as with any historical account, is better appreciated in the context of other events, and for this reason I have included the two tables below. The first provides a timetable for biological evolution, whereas the second provides a timetable for technological evolution. The dates are approximate; a two-fold uncertainty (although perhaps disturbing to the paleontologist and archaeologist) is inconsequential to the ensuing story of the human hand. But first let us peruse the tables.

One of the most striking features of the tables is their portrayal of a seemingly positive evolutionary momentum. For example, life first appeared on earth 3.5 billion years ago. It took more than 3 billion additional years for mammals to make their entry. Primitive mammals required 150 million years to evolve into the first primates (small, nocturnal creatures) which, in turn, required but 20 million years to become Old and New World monkeys. Only 3.5 million years span the entire human history from Australopithicus (the earliest known hominid and, basically, a two-footed ape) to modern man. In other words, humanoids occupy only 0.1% of the earth's entire biological history. Human technology has also advanced at an accelerating pace, taking 700 000 years to go from fire to the Sumarian tablets, and 3000 years to go from the Sumarian tablets to Shakespeare. Numbers like "3000" or even

"700 000" are miniscule relative to the evolutionary time-scale. One cannot gaze at the two tables (not to mention reflecting upon the technological progress of just the past two decades) without being struck how rapidly humans, and the products of their intelligence, have evolved.

Now back to the issue at hand. Prior to 10 million years ago, our ancestors were arboreal primates eating jungle fruit. (It has been claimed that our desire to eat a sweet dessert after a meal is a carryover from these times). In any event, the primate hands were designed for swinging from branch to branch. Then something remarkable happened: Bipedalism. We began to leave the forest canopy and occupy the African savannah on only two legs. Why we left the forest no one knows, but it might have been related to a climate change. It is far easier to formulate a Darwinian story for why bipedalism was advantageous. Perhaps we could jump higher to see game and predator animals in the tall grass. In the view of many, however, bipedalism did something more fundamental, namely it

Time-Table for Biological History (estimates)

Years Ago	Event
4.5 billion	Origin of the earth
3.5 billion	Origin of life
3.1 billion	First fossils (single cells)
500 million	First fish
400 million	First land planets and animals
300 million	First reptiles
200 million	First mammals
65–200 million	Age of dinosaurs
50 million	First primates
30 million	First Old and New World Monkeys
6–8 million	Human and chimp lines split
3.5 million	Austalopithecus
200 000 (?)	Anatomically modern humans with modern brains
60 000	Africa-to-Asia migration
35 000	Humans in Europe
15 000	Humans in the Americas

Time-Table for Technological Evolution (estimates)

Years Ago	Event
6–8 million	Bipedalism
2.5 million	First stone tools
700 000(?)	First use of fire
500 000	Hunting bands
150 000	Neanderthal tools
50 000–100 000(?)	Vocal language
40 000–50 000	Sophisticated tools, ornaments, cave art, etc.
10 000	Agriculture
3200 BC	First written language
2700 BC	Great pyramid
2000–3500 BC	Bronze Age
600 BC	Greek temple
1850 AD	Industrial revolution
20th Century	TV, penicillin, laser, atomic energy, auto, etc.

freed two limbs for other activities. After suitable modification, hands became available for the wielding of weapons; the transporting of meat back to camp; the carrying of babies from campsite to campsite; the manufacturing of tools. Such activities would have been closely associated with greater mental capacity, but that is not the whole story. Hands, along with communication skills, allowed a uniquely human lifestyle in which our young must be attended to for 16 years or more, a time period needed by our special brains to reach maturity. No other animal comes close to this requirement. Prolonged parenting must have been a terrible burden on the prehistoric family. For example, a female would have had to watch over her child continuously, for a span of years, and thus be unable to help materially with the hunts. One can imagine the nuisance of carrying or dragging a 40 pound child for long distances as the game animals migrated. Fortunately, hands with their opposed thumbs and nimble fingers, controlled by a smart brain, were present to help save us from extinction.

Another explanation for bipedalism has been advanced by Peter Wheeler. He proposed that standing upright reduced exposure to the remorseless sunlight of the African savannah. In other words, since our arboreal ancestors were adapted to the shade of the forest, the tropical sun in the treeless savannah could have presented a health threat that was partially alleviated by standing upright. Another recent theory has it that bipedalism actually evolved, or began evolving, when we were tree dwellers. Orangutans, who spend the majority of their time in trees, show a tendency toward two-leggedness when walking on the branches. Whatever one's favorite Darwinian story, the hands became free to do new and useful things.

In summary, the classical natural selection mechanism would go something like this: Our arboreal ancestors entered the savannah. Some of these, perhaps with a genetic predisposition for a bipedal terrestrial gait, began walking upright in which they could see further, or carry game, or wield weapons, or transport babies, or avoid the sun. Since these fortunate few would leave more offspring than their more quadrupedal counterparts, the upright trait increased in the population until everyone was walking vertically. It all sounds seductively reasonable and, no doubt, the model does have elements of validity. But then one begins to wonder about the details (those terrible details!): Why would a predisposition for bipedalism have developed in the arboreal primates? And if a predisposition for bipedalism had not existed, then how long would it have taken to acquire such a trait in a small group of slowly reproducing ancestors lumbering around the savannahs of Africa on four limbs? And when the multigene bipedalism trait did finally appear, how is it that the mutations "stuck", that is to say were imparted to the entire human population spread over a huge area?

Most sources, including this book, freely bandy about the term "hunter-gatherer" to describe the lifestyle of hominids prior to agriculture. Actually, there is a surprising amount of controversy even with this simple descriptor. The first stone tools date back to two million years ago and, from their earliest times, the tools were often associated with concentrations of animal bones. Many have deduced that cooperative hunting and sharing of meat had occurred, and that human evolution is closely related to the organizational demands involved with this cooperative effort. Others, however, have countered that hominids used the tools mainly for scraping meat and crushing bones at carnivores' kill sites, a possibility that has fewer social implications. Thus, current hypotheses have downplayed hunting in early hominid subsistence. "Man the hunter" has been replaced with "man the scavenger" or "woman the gatherer". A third model postulates that plants, not meat, supplied most of the food for early man. In support of this idea, micro-wear patterns on the surface of hominid teeth indicate that we were predominantly vegetarians until at least 1.5 million years ago. The hunting/scavenging/gathering debate emphasizes how even the most elementary aspects of our prehistoric past are not easy to decipher.

What, in summary, can be said about the origins of our intelligence? The sad fact is that we can say very little. We are reasonably certain that modern people with reorganized brains and speech abilities appeared on the scene in Africa roughly 50 000 years ago and then spread in two (or possibly more) waves into the rest of the world. There is little evidence that we have changed much genetically since then. Whether Neanderthal and even earlier man were equally smart, but only culturally backward, we do not know because there is no hard evidence one way or the other. Prehistoric brains do not survive fossilization, and even if they did it is questionable that one could learn much from them about neural rewiring. Information from tools is intimately tied up with culture. Tools, artwork, ceremonial items, etc. provide us with only suggestive evidence that indicates a rather abrupt appearance of modern intelligence a few tens of thousands of years in the past. But this flies in the face of Darwinism which says that organs of perfection evolved in tiny-step-by-tiny-steps over vast time periods. And gradual or not, Darwinism does not explain how hunter-gatherers might have given rise to an organ that houses a center for written vowels; that allows us to grasp the tenets of quantum mechanics; that controls piano and violin playing; that directs the writing of this book. We are certain that human nature developed as a coevolution of genes and culture, but other than this bit of "*obviousity*" (to coin a word), the whole business is very perplexing.

Chapter 8

POPULATION

Microevolution is a process whereby a single mutation, leading to a single new enzyme, modifies an organism. The development of bacterial resistance to antibiotics, such as streptomycin and penicillin, is a prime example and one simple enough to allow "back of the envelope" calculations. Consider a few streptomycin-sensitive bacteria on a culture medium that divide and produce, say, two billion (2×10^9) bacteria in a few hours. Spontaneous mutations to form streptomycin-resistance occur at rates of about one in 100 million (one in 10^8) cells. It follows that the bacterial culture will on the average generate (2×10^9) times (1×10^{-8}) or 20 streptomycin-resistant cells. If streptomycin is added to the culture, only these 20 will survive. The survivors will then start dividing and, in several hours, produce a colony of billions of bacteria, all resistant to streptomycin. This is microevolution in action.

Although the probability of finding a penicillin-resistant mutation is 1×10^{-8}, and the probability of finding a streptomycin-resistant mutation is also 1×10^{-8}, the probability of finding a single bacterium that has both mutations is (1×10^{-8}) times (1×10^{-8}) = 1×10^{-16}, an extremely small number. To obtain 20 such double-resistant bacteria in a few hours, one would have to prepare an absurd 100 million cultures. People have incorrectly deduced from such numbers that macroevolution (i.e. formation of organs, etc. via a series of microevolutionary events) is statistically unrealistic. The error in the reasoning lies in the neglect of selection. By doing the experiment stepwise (i.e. first selectively producing a large colony of streptomycin-resistant bacteria from the original 20, and only then repeating the experiment with a search for penicillin-resistance), one can reestablish the 1×10^{-8} odds per step. The key here is to first have a colony of two billion streptomycin-resistant bacteria before the penicillin is ever introduced. Selection and proliferation of the first mutation makes all the difference.

The preceding exercise was presented for one main reason. It illustrates how a huge population can counteract the effect of unlikely mutations. Although the streptomycin resistance appears only once in every 10^8 cells, having 2×10^9 cells makes the event likely. Population, in other words, is an important parameter in evolution, and it is therefore of interest to examine human population trends in prehistoric times. Large population would be conducive to rapid evolution because they provide an expanded pool of genetic variation from which the most reproductively advantaged members can emerge. Large populations also provide more stringent intra-species competition for food, shelter, mates, etc. and therefore, as Darwin fully recognized, a more ruthless selection.

Skeletal remains of prehistoric man provide a fair amount of demographic information. One can, for example, deduce the age of human skeletons at a campsite from the degree of tooth eruption, closure of cranial structures, and ossification of wrist bones. Estimating worldwide populations is more difficult, but reasonable guesses from the literature are listed below:

Years Ago	Population
2 000 000	125 000
300 000	1 000 000
25 000	3 000 000
8000	86 000 000
250	728 000 000

Not included in the table is a recent estimate that only a handful of modern humans (perhaps 5000) left northeast Africa to occupy

Europe and the rest of the world. We are, at least according to this theory, all descended from a mere 5000 migratory people living roughly 50 000 years.

Even late in our evolution, only 15 000 years ago, most of the planet was occupied by humans, but our population density was very low. Huge land masses were only sparsely populated with migrating bands of 25–50 people engaged in hunting and gathering. As seen from the population spurt 8000 years ago, it was only the advent of agriculture that allowed humans to expand their numbers dramatically. By then the majority of our genetic makeup was seemingly in place. One can hardly invoke bacteria, whose billions of rapidly reproducing numbers favor single-gene microevolution, as a model for human multiple-gene macroevolution. The numbers game was definitely not in our favor.

One of the most striking documented demographic facts of early man is their small family size. The number of children was no doubt limited by the need for a woman to break camp frequently and transport her baby and possessions to better hunting or food gathering grounds. A second child would hardly have been welcomed when survival depended upon mobility. How was family size kept small? The answer can be subdivided into two categories: "natural limitations" and "self-imposed limitations".

Natural Limitations

a) Life expectancy was low in prehistoric times, with skeletal evidence indicating that perhaps 50% of the population died before the age of 14. Survival beyond 40 was rare. Typically, a woman would begin reproducing after a late menarche, estimated to have occurred at the age of 16 or 17, until her death at 25–30. Thus, short lifetime, late menarche, long gestation period, and single-child births all restricted family size and population growth.
b) Disease and parasites would have taken their toll both in terms of fertility and infant mortality. Malaria, for example, is thought to have afflicted man from the earliest times.

c) Stillbirths associated with congenital defects might have been common within the small inbreeding hunting bands.
d) Cold and damp were deadly enemies to young and old alike.
e) Malnutrition from fluctuating food supplies probably limited fertility and family size.

Self-Imposed Limitations

a) Intervals between pregnancies might have been deliberately spaced out by prolonged breast-feeding.
b) Prehistoric people probably had few compunctions against engaging in infanticide given that the practice has been common throughout history. Thus, some Australian tribes killed every child born before its elder sibling could walk. Among the Pima Indians, a child born after the death of its father was killed so that the widowed mother would not have the extra burden. Regular female infanticide was at one time practiced in places as diverse as Tahiti, Taiwan, North Africa, and India. Infanticide was widespread throughout the Roman period although technically it was a crime. The Roman historian Tacitus found it foolish of the Jews not to condone it.
c) We do not know (and can never know) whether prehistoric humans limited their family size via contraceptive methods, but there is no doubt that the ancient world had available a rather impressive contraceptive technology. An anti-fertility tea made of the herb *Lithospermium ruderale* was taken by the Nevada Indians. Of course, many such potions (sometimes concocted symbolically from fruitless plants) were ineffective, as were measures, such as recommended by the Greek physician Aëtios, to apply vinegar to the penis. But the ancient use of the pessary (a vaginal suppository that kills the sperm or blocks their path through the cervix) might indeed have worked, ancient pessary recipes having been recently tested with success. The Petrie Papyrus of 1850 BC (one of the oldest papyri in existence) contains three prescriptions for vaginal pessaries. One is based on crocodile dung, another on a mixture of honey and sodium

> carbonate, and the third on a natural gum. Their consistency at body temperature was designed to cover the cervix with an impenetrable coating. Anthropologists in Sumatra encountered a traditional suppository which, it turned out, contained a spermicide: tannic acid. Women of Africa, Easter Island, and Japan have used plugs of chopped grass, seaweed, and bamboo tissue, respectively. In the early 20th century, women of the Djuka tribe of Surinam inserted a 5-inch long okra-like seedpod from which one end was cut off.

Perhaps methods such as the above were also known to the prehistoric woman. Or perhaps Nature was so harsh that no special care had to be taken to limit birth. Whatever the case, the prehistoric family was small. And, prior to agriculture, the total human population, scattered all over the world, was less than the population of many modern cities.

Let me now, in summary, list the salient facts about human prehistory and demographics: Early man, in the form of *Homo erectus*, evolved in Africa 1.6 million years ago. About one million years ago, there began a migration out of Africa into Eurasia from where humans spread all over the world. Until the dawn of civilization 10 000 years ago, with its agriculture and community life, man survived by hunting and gathering. The skills manifested by the hunter-gatherers improved with time. Thus, the Neanderthals of 200 000 years ago had only crude stone tools and may not have been able to speak. Cro-Magnon appeared 40 000 years ago with more sophisticated tools, artwork, and an ability to communicate through speech. The hunting-gathering period probably never reached a population in excess of ten million people.

How then did intelligence arise? The question is easy to answer: We do not know! But we do know that innate intelligence is universal; that no one group of people has a monopoly on intelligence, a fact consistent with a remarkably small genetic variation throughout the human population. At least two possibilities present themselves by way of explaining the universality of intelligence: (a) Intelligence might have been already present in our ancestors at the time they left

Africa and, by this means, the trait was given to everyone as the earth was populated. (b) Intelligence might have appeared in one group (e.g. the Cro-Magnon) and then spread, hunting band-to-hunting band, to all peoples. Both rationales have serious difficulties. In the first one, we are forced to postulate that our earliest ancestors, more ape than man, had an inexplicable intelligence on par with that of today (the magnitude of which I have already expounded on at great length). In the second rationale, we are forced to postulate an efficient genetic mixing among a sparse population spread across a huge land mass, covered with jungle, forest, sea, mountains, and steppes, with walking as the sole means of transportation. Population geneticist Shahani Rouhani calculates that it would have required almost half a million years for a single advantageous gene to travel from South Africa to China by the normal process of gene flow. And I am not talking about a single advantageous gene but a whole family of intelligence genes appeared and had to spread independently.

There is a school of thought, it must be pointed out, that believes that small populations are more conducive to evolutionary change. The argument is that small populations escape the "inertia" of large populations. Consider, for example, a small group of birds that possess genetic elements that are present only in low frequency among a much larger population. If birds from the genetically unique group happen to populate a remote island, then the originally rare genetic elements will likely assert themselves within the colonizing population (a so-called "founders effect"). Genetic drift, a related source of diversity, also takes place within small and isolated groups. If intelligence developed, in part, by one of these "small population" mechanisms, then one is pressed to explain how a trait arising in, say, a remote mountain valley, with little outside contact, became disseminated universally among each and every human being worldwide.

As was seen in the creation of antibiotic-resistant bacteria (described in the beginning of the chapter), two factors favor microevolutionary change: high population and high mutation rate. Any notion of a large human population during evolutionary times has just been dispelled. A high mutation rate is also unlikely, our DNA now being more than four times resistant to change than that of a mouse. Should there be doubt as to the stability of our genome, consider the low-to-negligible incidence of harmful mutations among the children of men over 50 years old (i.e. men who have been subjected to mutational effects in the environment for over half a century). Or consider the children of Hiroshima and Nagasaki. Contrary to general belief, children conceived after the bombings showed little or no deleterious effects. Apparently, the radiation damage to the germ cells was either (a) drastic and thus eliminated as spontaneous abortions or (b) less severe and correctable by our efficient DNA repair systems. The net result was a rather normal population of children, indicating again that the human genome is surprisingly resistant to mutational damage by radiation.

Anyone writing about intelligence does so from a position of ignorance; we simply lack the necessary genetic and embryological information. For example, no one knows how our DNA sequences, producing mere proteins, translate into complex neural connections associated with brain function. Nor do we even know how many genes affect the operation of the brain. Since eye color in the *Drosophila* fruit fly can be modified by 14 genes, it is a safe guess that far greater numbers must play a role in human intelligence. After all, the human brain is an "organ of perfection".

To summarize in one sentence much of what has been argued thus far in this book:

> *Natural selection is faced with the problem of explaining how a* ***complex set of genes****, controlling an* ***expensive trait*** *with no obvious benefit, came into* ***permanent existence*** *in such a* ***short time period*** *within* ***every member*** *of a* ***small population*** *(that was* ***dispersed and geographically isolated*** *over the entire planet) who had a* ***low reproductive output*** *and a* ***low mutation rate****.*

A brief expansion of the parameters in the preceding statement will help the reader grasp the magnitude of the dilemma:

a) "***complex set of genes***": Fourteen genes control the eye color of fruit flies. Many hundreds of genes affect human odor detection. It is not known how many genes must operate in concert to create human intelligence, but it must be a large number.

b) "***expensive trait***": About 20% of our energy consumption is devoted to the brain.
c) "***no obvious benefit***": Written vowel centers in the brain, solving differential equations, and playing the violin were, for example, of no obvious benefit to hunter-gatherers.
d) "***permanent existence***": Our permanent acquisition of intelligence is obvious.
e) "***short time period***": Modern humans (Cro-Magnons) appeared only about 40 000 years ago. It is not clear whether humanoids prior to the Cro-Magnons even had the vocal apparatus necessary to speak.
f) "***every member***": Uniform intelligence distributions are found throughout the world. Whatever mechanism is proposed for the trait's evolution, the mechanism must explain a "perfect mixing".
g) "***small population***": It is estimated that 25 000 years ago the world population was only about 3 million.
h) "***geographically isolated***": Distributing a mere 3 million humans across all continents (except the Americas and Antarctica) must obviously have limited reproductive interactions and the "perfect mixing" of intelligence genes.
i) "***low reproductive rate***": According to fossil records, perhaps 50% of the population was dead before the age of 14. Disease, starvation, cold weather, homicidal raids, and infanticide took their toll.
k) "***low mutation rate***": Humans are generally harmed, not benefited, by mutational assaults. X-ray-induced mutations, for example, often lead to cancers and other maladies. Even if a rare beneficial mutation does appear, it can be removed by well-known repair mechanisms, thus limiting the potential for Darwinian evolution.

None of the above factors, be they genetic or demographic, is favorable to the furtherance of natural selection. We must search elsewhere for an explanation of why humans are so smart.

Developing ideas via "book format" has a peculiar and unavoidable difficulty: Readers quite naturally tend to forget, or at least set aside, important points in early chapters by the time concepts in later chapters are expressed. Thus, to fully appreciate the significance of my summary sentence, it might be helpful to refresh your memory of the sections on linguistic, musical, and mathematical intelligence. Then return to the above sentence where, using a few phrases (in bold), I summarize the current evolutionary dilemma as succinctly as I know how.

Chapter 9

CULTURE

There are over 20 000 genes in the human genome (a number that has been adjusted downward from estimates as large as 100 000 in the recent past). Let us accept ball-park guesses of one trillion nerve cells in the human brain engaged in about 300 trillion connections (synapses). This means that there are about at least 1.5 billion synapses per gene (the number being a gross underestimate if one considers the fact that most genes are not engaged in neurological activities). Even using an overly conservative guess of 30 trillion synapses, one calculates 150 million synapses per gene. Genes may give us our neurons and synapses (in ways that are not understood owing to our primitive understanding of developmental biology), but genes do not control most of the wiring of the resulting network. Networks that are ultimately established in the brain of an adult human must, consequently, be largely determined by environmental and cultural factors. But we all know this. After all, a child is not born speaking French or solving an equation. Genes provide the hardware, culture the software, and reality is created by the combination and overlap of the two. As John Allman has stated, "the brain is unique among the organs of the body in requiring a great deal of feedback from experience to develop its full capacities."

Now it may seem, at first thought at least, that the cultural component of intelligence solves, or at least addresses, the dilemma of explaining the evolution of human intelligence. Thus, one could argue that modern humans appear so smart because they are subjected to culture e.g. schooling, music lessons, computer games, map reading, and innumerable other intellectual environmental inputs. Although this is a correct statement, it only adds to the mystery of how our smartness evolved. Thus, for most of our evolutionary experience, we did not encounter book learning, essay writing, equation solving, violin playing, or multiple languages. Yet we possess, universally, the hardware to master all of these through the educational process. Clearly, our brains have the capacity and flexibility to accomplish difficult activities that have never had a parallel in the long story of our cultural background. How could essay writing, for example, ever had evolutionary survival value when no one could write until very recently? How can one invoke culture as a source of literacy when we evolved as illiterate hunter-gatherers?

Somehow or the other, human brains evolved to carry out exploits far in excess of what was needed at the time. Even today, it is claimed, we utilize only a fraction of our 300 trillion synapses. Human brains seem to have "overdeveloped" by some miraculous quirk of fate that we do not understand. Natural selection, whether genetic or cultural, does not provide insight into the problem because an expensive "overdevelopment" lies outside the domain of the theory. Natural selection, with all its virtues and power, seems inadequate to serve as the sole mechanism for the evolution of the organ that, more than any other, makes us human.

The preceding is not to suggest that culture cannot alter gene populations. It is obvious from dog breeding programs that human cultural practices can affect genetic makeup in animals. The development of vision-correcting eyeglasses no doubt aided the spread of myopia genes that previously must have experienced significant negative selection pressure. Dairy farming is believed to have increased the fraction of the population that produces a gene, and corresponding enzyme, capable of breaking down milk sugar. And (who knows?) perhaps sexual selection among early humans favored those who were good storytellers with good memories and communication skills. Cultural evolution itself is different from genetic evolution in that only the former is fast (a widespread fad can appear overnight), can be passed along among relatives and non-relatives alike, and is intragenerational.

The key question is whether our higher intellectual faculties have evolved as an adaptation to the complexities of social life. In other words, one wonders if human nature, including intelligence, has been shaped by social interactions (friendly or otherwise) among family members, people in the band or village, and various outsiders. Since many cultural anthropologists and sociologists believe that the origin of human intelligence lies in social intercourse, aided by the use of communication through speech, serious consideration must be given to this possibility.

Serious consideration will not be given, I add parenthetically, to Peter Dawkins' inexplicably popular notion of a "meme" as a unit of culture. The meme, intended to be the cultural counterpart of the gene, is arbitrary and vacuous (as can be seen if one fruitlessly attempts, for example, to define a specific meme for music or mathematics in today's culture).

I should also add parenthetically that some people, most notably William Rathbone Greg, a 19th century Scottish moralist and political writer, considered cultural natural selection as actually counter-evolutionary. The argument is that sympathetic societies protect the physically, intellectually, and morally inferior from the hand of natural selection. Since these latter folks, according to Greg, procreate faster than their protectors, society would deteriorate rather than advance as a result of cultural practices. Eugenics, the discredited idea of improving humans through controlled reproductive schemes, is closely allied to Greg's philosophy.

Far more worthwhile is the idea that intelligence arose as a non-adaptive side effect (a "*spandrel*", in the words of Gould and Lewontin, but also referred to as an "*exaptation*"). This interesting idea argues, correctly, that adaptive changes necessarily produce genetic by-products that might or might not be later co-opted for useful purposes. The prevalence of spandrels is beyond doubt. For example, mammalian blood is red because the hemoglobin in the blood, which carries oxygen, happens to be red. But the red color trait is not adaptive; it has no useful purpose per se; its presence is an accidental by-product. Red blood, being no more than a side effect of our need for hemoglobin, is therefore a good example of a spandrel.

Yet it is difficult to believe that most of our mental properties (such as reading, writing, and mathematical ability) are mere spandrels associated with other unidentifiable traits that did indeed have substantial impact on evolution. If this were true, then the most unique of human traits, intelligence, is merely a wondrous accident...a serendipitous event that lies outside the province of natural selection theory. Worse, the spandrel concept, when applied to intelligence, actually runs counter to natural selection. While natural selection has no problem accepting and explaining the red color of blood (this spandrel being produced as a cost-free by-product of the hemoglobin gene), natural selection cannot embody a trait such as intelligence that arises, and is maintained through the ages, as a complex, useless, non-adaptive, and a high energy-cost side effect.

As always, it is possible to construct an evolutionary story. I will now attempt to do so with a story that embraces a spandrel-based intelligence source: Owing to certain genetic changes, a prehistoric male ancestor of ours, living in Africa 150 000 years ago, became endowed with a greater number and viability of sperm. Simultaneously, these same genetic changes provided the man, as a spandrel, a neural circuit that allowed him to write (a useless ability, of course, because cultural progress had not yet incorporated writing skills). Owing to the greater fertility of the ancestor, his genes were propagated at an above average rate, so that ultimately the writing trait, although totally non-adaptive, also became prevalent in the population. Although the writing trait was unutilized and energy-expensive, the high sperm count "covered the cost", evolutionarily speaking. Then, 150 000 years later, when writing became a component of our culture, the genetic side effect suddenly became a useful thing to possess.

Unfalsifiable, post hoc storytelling would be an appropriate descriptor. For a starter, Mother Nature does not foresee a future need for a trait. And it is impossible to argue convincingly, even by the most skillful evolutionary storyteller, that genes controlling sperm count can, as a peripheral by-product, also control writing centers in the brain. In other words, we are asking one single complex set of genes to dictate the formation of two entirely unrelated traits…a highly unlikely scenario.

Now let us return to the more plausible idea that cultural pressures to compete and cooperate were the driving force behind human intellectual development. In other words, the neural circuits used today to solve complex problems evolved from neural circuits used originally by ancient humans in order to deal with one another and with the environment, whether it was organizing a hunt, attracting a mate, teaching the children, fighting off enemies, telling a story, sharing a kill, or exchanging information. Groups with a greater number of "cooperation genes" (if such exist) would ostensibly have spent less time arguing and wasting time that could be spent more profitably with useful activities such as hunting game. Natural selection pressure would therefore have favored the proliferation of these genes. Culture (including sexual selection) and genetics are thereby intertwined, a fact that gradually gave rise, according to the theory, to modern intelligence. It is a seductively reasonable (although somewhat nebulous) explanation for why humans are so bright.

What are the problems with the assertion that cultural factors in the distant past, including language development, led directly to the intelligence familiar to everyone today? To a large extent, the question has been already addressed in the sections on language, musical, and mathematical intelligence. The difficulty of explaining the spread of intelligence genes, culturally related or not, to all humans (and all humans are endowed equally with them), considering that prehistoric humans were living in small bands spread thinly across vast land masses, was discussed in the previous chapter on Population.

Humans have no doubt evolved intelligences specialized for social interactions, but these intelligences are not identical to those for violin playing or equation solving (called "intellectual intelligences" for want of a better term). Support for this statement is persuasive: Individuals with autism can lose social intelligence while maintaining a high level of intellectual intelligence. There is no evidence that people nowadays (and presumably in the past) with outstanding intellectual intelligence (e.g. world-class philosophers, physicists, composers, etc.) are more reproductively successful than anyone else. Indeed, the opposite may be true. Intellectual boys, often stereotyped as being "nerdy", are generally not as popular with the girls as are the macho football types. Teenagers attain emotional highs at rock music concerts, not at poetry readings or chemistry classes. College cheerleaders shake, dance, and flip (akin to a modern puberty rite) at basketball games, not theatrical or orchestral performances. Nietzsche thought that the mere idea of a married philosopher was comical. And it is the case that few CEOs of major corporations are scientists, folks with a high level of intellectual intelligence; most CEOs are drawn from advertising, marketing, legal and other non-technical circles who have a high degree of social skills and intelligences. Thus, the argument that a prehistoric human mutated to develop social skills, and that these skills were genetically allied with non-adaptive skills such as literacy and violin playing, and that the neural equipment appearing in this individual, wherever he or she lived, was passed on uniformly to every human on earth via a process of natural selection over a relatively short time period...well, something is missing, something does not quite add up. In the words of linguist Steven Pinker, "The apparent evolutionary uselessness of human intelligence is a central problem of psychology, biology, and the scientific worldview."

I previously introduced the term "plasticity". This term refers to the absolutely amazing capacity of the human brain to adjust to external conditions. For example, if you lose a finger, the area in the brain controlling that finger shrinks, while the area controlling the other fingers expands to take its place. There is evidence that the visual cortex of people blind at birth does not wither away but, instead, takes on new tasks such as hearing or visualizing textures felt on the fingertips. Strokes that destroy a blind person's visual cortex can also destroy the person's ability to read Braille with the fingers. I have already suggested that plasticity is an important component of intelligence and, to some degree, intelligence can also be equated with plasticity.

Neuroscientist Mriganka Sur redirected the visual connection in newborn ferrets from the visual cortex to the auditory cortex. Thereupon, the auditory cortex became capable of processing visual input from the eyes. Thus, genes do not predetermine cortical function. Instead, genes create a plasticity that, upon further developmental and cultural refinement, produces the end product.

We do not know this, but it is quite possible that spear point manufacturing, and its attendant visual and manual inputs, helped reprogram and redirect the development of the plastic brain parts controlling vision and the hand. It is no surprise that neural plasticity, and the corresponding potential for the brain to do more than it had evolved to do, has been invoked to help explain the origin of human intelligence. But, once again, we are confronted with a problem:

A plastic brain neurologically altered by an external stimuli, such as a missing finger, or blindness, or tool making, is not a heritable trait. It is an acquired trait. If current anti-Lamarckian arguments are to be believed, acquired traits cannot be passed on to the progeny except by cultural routes (i.e. teaching). Teaching is no doubt a rapid and effective way to pass on a cultural innovation, but the innovation must be introduced to each new generation because it is not inherited. How, one asks, did humans acquire their amazing neural plasticity capable of responding to challenges, such as literacy, never before encountered? In this sense, intelligence and plasticity entail an equivalent problem with regard to understanding evolution in terms of natural selection. One is reminded here of a theory that life itself began when a meteor, bearing primitive life, landed on earth. Right or wrong, the theory provides little fundamental understanding because the essential problem remains; we must still determine how life formed in outer space. The origin of life question has not been solved, only displaced. And so it is with neural plasticity. Neural plasticity does not solve the problem of intelligence; it only displaces it. Evolution inexplicably provided *Homo sapiens sapiens* with an expensive supercomputer at a time when programming to fully operate the supercomputer was not available.

There are those who argue that the brain of *Homo sapiens* has not changed for 100 000 or even a million years. This theory has its appeal. It means that humans had attained current intelligence levels well before we ever left Africa. Human progress, therefore, was (a) a matter of discovering uses for the vast neural networks that were already on hand, and (b) passing the ensuing capabilities on to the next generation through education...in other words, culture. Thus, prehistoric development, once we were out of Africa, was a result of cultural innovation, not mutational development of cranial abilities. According to this viewpoint, culture simply promoted the use of a brain that became fixed (perhaps 150 000 or more years ago; no one really knows) at modern levels of intelligence.

Social learning, in contrast to genetic factors, can spread innovations fairly rapidly. This represents an important advantage of the cultural theory of intelligence over the genetic theory. It is easy to imagine how a cultural development (e.g. pottery making) in one population group can spread rapidly to other populations. A Japanese experiment demonstrates that such spreading of information would not have even required an ability to speak: Undergraduate students were divided into two groups. One group was taught how to make a typical Neanderthal stone tool using verbal explanations and demonstrations. The other group was taught by silent example alone. There was no difference in the speed or in which the two groups acquired the tool-making skills nor in the quality of the tools. Learning by silent example worked just fine.

The cultural stimulus theory solves another dilemma, namely why all peoples of the world are inherently equivalent in their mean intelligence. They are equally bright because they all descended from a common African ancestor who possessed a modern intelligence. We no longer have to worry about how all those intelligence genes (appearing randomly and scattered among people living in, for example, the Russian steppes, the Australian outback, and northern Europe) managed to integrate in a short time period and thereby endow intelligence uniformly among all populations on earth.

Unfortunately, the "out of Africa theory" has its own serious problem: How can one explain by natural selection the abrupt quantitative and qualitative leap to modern intelligence (among African ancestors living not too differently from apes)? Natural selection works by minute genetic changes in tune with the environment, not by bursts of under-utilized complexity. Either way, within Africa or outside Africa, natural selection poses serious questions about the origins of intelligence that cannot be answered at the present time.

Chapter 10

ANIMAL INTELLIGENCE

In the "*Descent of Man*", Darwin wrote: "There can be no doubt that the difference between the mind of the lowest man [sic] and that of the highest animal is immense." He went on to write (in that wonderful Darwinian style): "An anthropomorphous ape, if he could take a dispassionate view of his own case, would admit that though he could form an artful plan to plunder the garden, though he could use stones for fighting or for breaking open nuts, yet that the thought of fashioning a stone into a tool was quite beyond his scope. Still less, as he would admit, could he follow out a train of metaphysical reasoning, or solve a mathematics problem, or reflect on God, or admire a grand natural scene." "Nevertheless," Darwin continued later, "the difference in mind between man and the higher animals, great as it is, certainly is one of degree and not of kind." This last assertion engendered a debate that has persisted till present times: Is human intelligence superior to that of animals in degree or kind? That is the sort of question that scientists would love to ask because, in the absence of definitions of "degree" and "kind", there is no possible resolution and, therefore, no end to the journal articles in which scientists can expound on the subject. My own view is that, in agreement with Darwin, the difference between man and animal is immense, and I leave it dangling there without a hard definition of "immense".

I want to include a short chapter on animal intelligence because I must admit to another purpose in writing this book other than to discuss natural selection, human intelligence, and the relationship between the two. I want to illuminate all of Nature. Whether by natural selection or some other mechanism, Nature has blessed us with diverse animals that, although not as intelligent as ourselves, are, each in its own way, magnificent. I do not want my book, emphasizing as I do the mysterious power of human intelligence, to leave the impression that we are superior to earth's other denizens. And if we have any sense (which is quite different from intelligence), we will ensure that the earth provides suitable homes for these other animals. Who wants to live in a world occupied only by crowded humans, their pets, and their livestock? Who wants to visit the mountains and not see other animals, an experience that would treat us to scenery but not wilderness? I admit that most people (including myself) have never seen, and will never see, a tiger in the wild. Yet if I am ever told that tigers in the wild exist no more, I will be diminished. Something in my "thin bone vault", I do not know exactly what, will cause me to feel revulsion. I therefore dedicate this chapter to the preservation of those wonderful wild animals from which we humans, smart as we are, can learn a great deal.

Let me begin with the famous case of Japanese macaques (a large monkey). In an attempt to relieve hunger in an overpopulated community of macaques on an island in southern Japan, primatologists threw gains of wheat upon a sandy beach. The macaques were anxious enough to eat the wheat grains, but they had difficulty separating the wheat from the sand. One day a macaque named Imo did a brilliant thing: She threw handfuls of sand plus wheat into the water. Imo noted that the sand sank, whereas the wheat remained on the water surface where she could easily scoop it up. Although the older macaques ignored the discovery, the younger monkeys (and this is amazing) appeared to grasp the importance of the new technology and to imitate it. The practice became more widespread with each generation until today all the macaques on the island are competent at "water sifting".

While on the subject of non-human primates, I might offer two more quotes, the first coming from Thomas Huxley (a contemporary of Darwin and one of his staunchest supporters) who compared humans and apes as follows: "No one is more strongly convinced than I am of the vastness of the gulf between…man and the brutes for he alone possesses the marvelous endowment of intelligible and rational speech [and]…stands raised upon it as on a mountain top, far above the level of his humble fellows, and transfigured from his grosser nature by reflecting, here and there, a ray from the infinite source of truth." My main quarrel with the quote is not so much with the sentiment as with the use of the word "brute", a word that implies a gross, vile, irrational, and depraved animal. Imo could not be categorized as any of these.

In a closing passage of "*Descent of Man*", Darwin compares the monkey to humans living at the tip of South America, and the monkey actually gets the better of the comparison: "The main conclusion arrived at in this work, namely that man is descended from some lowly organized form, will, I regret to think, be highly distasteful to many. But there can hardly be a doubt that we descended from barbarians. The astonishment which I felt on first seeing a party of Fuegians on a wild and broken shore will never be forgotten by me, for the reflection at once rushes into my mind — such were our ancestors. These men were absolutely naked and bedaubed with paint, their long hair was tangled, their mouths frothed with excitement, and their expression was wild, startled, and distrustful. They possessed hardly any arts, and like wild animals lived on what they could catch; they had no government, and were merciless to every one not of their own small tribe. He who has seen a savage in his native land will not feel much shame, if forced to acknowledge that the blood of some more humble creature flows in his veins. For my own part I would as soon be descended from that heroic little monkey, who braved his dreadful enemy in order to save the life of his keeper, or from that old baboon, who descending from the mountains, carried away in triumph his young comrade from a crowd of astonished dogs — as from a savage who delights to torture his enemies, offers up bloody sacrifices, practices infanticide without remorse, treats his wives like slaves, knows no decency and is haunted by the grossest of superstitions."

Harold Bauer recounted the remarkable behavior of a male chimpanzee as he followed it through the jungles of the Gombe Stream Reserve in Tanzania. Deliberately or not (this cannot be determined), the chimpanzee came upon a 20-foot waterfall cascading down into a mist-enshrouded pool. The animal responded to the beautiful site with great excitement: calling out, rocking, running back and forth, jumping, and drumming his fist on the trees. This was not a one-time display; the chimpanzee returned days later and repeated the performance as did other chimpanzees on other occasions. No practical "animal" reason for the behavior was evident. The chimpanzee did not attempt to drink from the stream or to cross it (which, in any event, would have been easy to do had he so wished). Feeding, fleeing, fighting, mating, etc. were not factors here. Someone with an anthropomorphic bent might conclude — and who knows, perhaps correctly — that it was the scene's beauty that elicited the behavior. Perhaps emotions such as surprise and curiosity were also important. I could, to be argumentative, propose another explanation: The waterfall might have been the site of a past tragedy such as the killing of a chimpanzee by a predator. Could it be the memory of that event, or possibly some lingering odor, which caused the chimpanzees to be excited?

Jane Goodall, who also worked with chimpanzees in the Gombe Stream Reserve, described animal grief in an old female chimpanzee named Flo and in her 5-year-old son Flint. Flo was still nursing Flint but stopped when she became pregnant with Flame. Flint whined and moaned and threw tantrums at the lack of attention until, at the birth of Flame, his misbehavior improved somewhat. Flint in fact became quite solicitous of Flame. At the age of six months, Flame died of an apparent infection. Flo and her son appeared to take solace in each other with, for example, Flo again nursing Flint. Flint continued as a "mamma's boy" for two years thereafter when Flo suddenly died. Flint remained near the body in an obviously dejected state. He was so depressed that he did not feed or take care of himself although by this time he knew how to do so. A few days later Flint died too. Since an autopsy showed no apparent cause of death, one is tempted to assume that

Flint died of a broken heart. Of course, we do not know whether Flint grieved over a specific individual or for an absence of material things associated with that individual. But then again even human grief is a mixture of the two.

Grief behavior is species dependent. For example, an infant bonnet monkey responds to the removal of its mother with protests and sadness, but it also initiates affectionate relations with other adults, one of which usually adopts it. In contrast, a pigtail monkey that loses its mother makes no attempt to reestablish another adult relationship; it simple curls up into a little ball in the middle of the cage. The adults, in turn, make no effort to assist the infant. Whether the behavior difference is cultural, "hard-wired", or (most likely) a combination of both is not known.

Frans de Waal, in his book "*Good Natured: The Origins of Right and Wrong in Humans and Other Animals*", documented various chimpanzee emotions including compassion, sympathy, altruism, self-recognition, and even a sense of justice. Let me cite a few observations supporting the presence of such complex feelings:

a) Two adjacent cages with a common wire mesh wall were set up, one of which had within it a food "vending machine" operated by coins. Only the other cage had a set of coins. A chimpanzee trained in the use of the machine was placed in the cage with the machine. A chimpanzee with the coins in the other cage, passed the coins from cage-to-cage to his neighbor so that the latter could operate the vending machine.
b) Using the same adjacent cages as in the previous experiment, de Waal gave one capuchin monkey a bowl of apple slices, the other a bowl of cucumber slices. The monkeys handed, pushed, or threw their food through the wire partition so that they could, seemingly, share each other's bounty.
c) Bonobos, sometimes called pygmy chimpanzees, have been observed to play "blindman's bluff" with dedication and concentration. Thus, a bonobo covers her eyes with a banana leaf, an arm, or with two fingers. Handicapped in this manner, she stumbles around and bumps into others while negotiating a

climbing-frame. The game ends when she begins to lose her balance, whereupon she uncovers her eyes.

d) If a fight breaks out between two chimpanzees over food, minutes later the aggressor will quietly groom the chimp at whom he bared his teeth earlier in an apparent attempt at reconciliation. Bonobos often ease the tension after a fight by engaging in heterosexual and homosexual activities (a sort of "make love, not war" philosophy).

e) It is easy to discern when a chimpanzee is upset: It pouts, whimpers, yelps, begs with outstretched hands, or impatiently shakes both hands. Other chimpanzees will respond to this seemingly urgent request for consolation by hugging, touching, and grooming the distressed animal.

f) If chimp A picks bugs off chimp B, then hours later chimp B will be much more likely to let chimp A have some of his food. And both will be much more likely to share their food with chimp C if chimp C shared food with them earlier in the week.

What is one to make of all this? First of all, much of the behavior is, I presume, culturally mediated but fundamentally preprogrammed. Unfortunately, we know little about how DNA sequences (the basis of inheritance) ultimately lead to complex and integrated behaviors and activities. There has been some progress in the general area, however. For example, testosterone (a steroidal hormone) is associated with male aggression. And enough is known about brain chemistry that depressed behavior can now be treated with specific drugs. There are those, such as Francis Crick of DNA fame, who believe that most emotions — even human ambition and love — will someday be describable in chemical/neurological terms. Somehow I hope that the day is a long way off. In any event, my book's main objective is to contemplate human intelligence, not human behavior per se, so I will not pursue the latter topic further.

My digression into animal behavior was motivated, in part, by my fascination with animal life and, in part, by the frequent comparisons of animal intelligence with human intelligence. Since one can judge animal intelligence chiefly via animal behavior, animal behavior is a legitimate concern. Now with regard to the de Waal chimpanzee experiments, a word of caution is necessary. Observing what animals do is one thing; speculating on what the animals might be thinking is quite another. To drive the point home, consider the behavior of ants in a large colony where death is a frequent occurrence. Ants are known to pick up their deceased sisters and carry them to a refuse pile distant from the nest. "How clever," one might say, "the ants realize that decay presents a health hazard, and therefore they are careful to keep their home neat and tidy. Clearly, ants and humans think alike in this regard." In actual fact, there is a chemical called a "funeral pheromone", emitted by dead ants, that compels live ants to pick up the corpses and carry them away. Chemists have synthesized this chemical and placed a tiny drop of it on a live ant. As one might expect, ants picked up the startled pheromone-ladened ant and transported it to the refuse pile. The lesson here is that an animal may behave as would a human in a similar situation, but this does not necessarily mean that the animal is thinking like a human or, in fact, is thinking at all.

de Waal is, of course, aware of the dangers of over-interpreting animal behavior. Although a chimpanzee may have feelings and emotions, it cannot (that we know of) conceptualize the principles behind its behavior. Chimpanzees cannot define what is right, and why, and what is wrong, and why. In the words of de Waal, "animals are no moral philosophers". Be that as it may, it is probably simplistic to assume that a chimpanzee's sharing of food with a hungry companion is purely instinctive, while a human doing the same thing is exhibiting moral decency.

Primates are by no means the exclusive owners of intelligence and feelings. Captive Komodo monitors (a large lizard) easily learn to recognize different people (awareness). Elephants doing lumbering work in India obey two dozen commands (learning). Wild elephants cluster around a wounded member of the herd and try to help it to its feet (sympathy). If the wounded elephant dies, the others are slow to leave the corpse and sometimes lay branches over it (grief). African buffaloes wounded by a hunter will, instead of escaping, often lie in wait to ambush the hunter (revenge). Clark's nutcracker, a bird, stores about 32 000 pine seeds per year in about 10 000 caches (thriftiness), and months later the bird will recall where most of them are (memory). Ground squirrels have different warning calls for hawks, coyotes, or snakes (communication). Porpoises can invent new tricks to please the trainer and earn a fish (creativity). Pandas slide down a snowy slope, climb back up, and do it again (amusement). Mother cats, leading kittens on hunting expeditions, partially kill a prey and let the young finish the job (education). A species of bird uses a thorn to extract a grub from a tree cavity (tool use). Dogs growl, yelp, and whine (anger, joy, and pain).

In winding up this essay on animal vs. human intelligence, I should quote from both sides of the "in-degree" vs. "in-kind" debate. Stephen Gould is of the "in-degree only" school and writes:

> Educated people now accept the evolutionary continuity between human and apes. But we are so tied to our philosophical and religious heritage that we still seek a criterion for a strict division between our abilities and those of a chimpanzee.

But David Berlinski believes in a deep intellectual chasm between humans and animals:

> No distinction? Chimpanzees cannot read or write; they cannot paint, or compose music, or do mathematics; they do not dine and cannot cook; there is no record anywhere of their achievements; they are born; they live; they suffer, and they die.

He continues:

> One may insist, of course, that all this represents a difference merely of degree. Very well. Only a difference of degree separates man from the Canadian goose. Individuals of both species are capable of entering the air unaided and landing some distance from where they started.

And where, in summary, do I stand on this "in-degree" vs. "in-kind" issue? As mentioned in the beginning, I am reluctant to enter a debate that entails two categories that have not been carefully defined. I will say this (and in no way do I mean to deprecate animal life which I value dearly): Humans, and only humans, seem to have developed a mental capability far in excess of what was needed during their evolutionary past. And it is difficult to imagine why natural selection should have wrought such a remarkable brain when, for most of our existence, we lived not much differently than the aforementioned aborigines of Tierra del Fuego described by Darwin.

Section 3

EVOLUTIONARY POTENTIAL

Chapter 11

INTRODUCTION

Let us pause here to summarize the main elements of the book as presented thus far. **Section 1** covered the principles of Darwin's natural selection. Many of the objections to natural selection, raised frequently by others, were shown to be either inconclusive or outright incorrect. **Section 2** discussed elements of human intelligence, a trait that does indeed appear to conflict with natural selection. This does not mean that natural selection must be abandoned. It does mean that, since natural selection lacks explanatory power in at least one critically important component of the human condition, the theory should be modified, or at least expanded, despite its otherwise substantial utility. Of course, until such a time that a more encompassing model is proposed that overcomes Darwinism's weaknesses, Darwin is best retained in full. Obviously, I have thus far not yet proposed a more general theory. To do so requires that I now discuss elementary aspects of modern genetics. Once these principles are digested, the door is opened to fascinating examples from Nature (e.g. snapdragons, the immune system, etc.). As in the previous parts of the book, I attempt to present in **Section 3** the material in as concise and palatable manner as possible. But every reader has a choice: He or she can profitably stop here, being (I hope) satisfied with the insights into human evolution and intelligence that have already been brought forth. Or the reader can continue with the short remaining sections of the book in which I become somewhat more technical but, in the end, also more innovative.

In **Section 4**, the reader will be exposed to speculative theorizing that extends and complements, but does not displace, the great ideas of Darwin. Interesting questions are addressed, e.g. Is it possible (or mere neo-Lamarckian nonsense) that the human brain is, somehow, in communication with germ cells? If this is the case, then does educating one generation favorably impact on the inherent learning ability of future generations? Asked in another way, can the human race be getting smarter and smarter apart from classical Darwinian selection? The hope is that by addressing questions of this sort, we can arrive at a better understanding of human intelligence and humanity itself. Perhaps the most important point made in **Section 4** is that alternative explanations for evolution are possible and reasonable (albeit without direct proof at the moment). If taken seriously, this section will force us to disband current notions that natural selection must be accepted as the sole source of evolution because “it is the only game in town”.

Chapter 12

ELEMENTARY GENETICS

Each chromosome (of which each human cell has 46) is comprised of two long, double-stranded molecules called DNA. A DNA chain is a linear sequence of "bases" consisting of only four types: A, T, G, and C. Thus, a segment of a DNA molecule might read as ...G-G-G-A-A-G-G-C-A.... The previous section on Molecular Evolution reviews the topic. A human chromosome is 1.8 meters long with 6 × 109 bases (all compressed within the cell nucleus). Groups of a few hundred to several thousand bases along this DNA chain constitute a "gene". The rule that "one gene gives rise to one protein" is probably the most important generalization in biology:

sequences of bases in a gene
→ sequences of amino acids in a protein

Genes in the nucleus give the thousands of cellular proteins according to a complicated mechanism which need not be detailed here (with apologies to the many Nobel Prize winners and countless others who have worked on the DNA problem). It is sufficient to know that genetic information (i.e. A/T/C/G sequence in the genes) leads to precisely defined amino acid sequences in the proteins. And proteins are the essence of many life processes. For example, humans have an oxygen-carrying blood-protein, called hemoglobin, which is the product of a gene that specifically codes for it. Cows also have a gene for hemoglobin, but its amino acid sequence is slightly different from ours because the corresponding gene is not exactly identical.

Proteins are made up of sequences of (typically) 250 to 1000 amino acids long. Since there exist 20 different amino acids, the number of possible proteins is virtually infinite. Genes determine the amino acids sequence in proteins according to a "triplet code" in which three bases specify one of 20 amino acids (e.g. GGG = glycine, an amino acid; AAG = lysine, another amino acid; GCA = alanine, yet another amino acid). The base sequence mentioned above, ...G-G-G-A-A-G-G-C-A..., would, therefore, lead to a glycine/lysine/alanine segment somewhere within a protein.

All in all, there is a rough correlation between the quantity of an organism's genetic material and its complexity. For example, it is estimated that the bacterium *E. coli* has 4288 genes, whereas humans have at least 20 000 genes. Before getting too proud of our genome, however, we should realize that the DNA content of the lungfish is 10–15 times larger than ours, and certain algae can have genomes an order of magnitude larger than this. Rough guesses give 250 genes as the minimum number required to sustain life. Most of these are devoted to the production of critical enzymes (an enzyme being a protein that catalyze biological reactions).

Consider the consequences of a "point mutation" in which AAG, the gene code for the amino acid lysine, converts into a different triplet, GAG. Since GAG is the code for the amino acid glutamate, the result of this A-to-G mutation would be a protein in which a lysine has been replaced by glutamate. In terms of biological activity, the replacement could improve the protein but, much more likely, it would create a defective protein. As one might expect, most random mutations in a gene's DNA are harmful to the organism because they cause a modification of a delicate protein structure that has been evolving toward perfection for millions of years. Random tinkering with a watch, to give again an oft-cited analogy, seldom does the watch any good.

In the words of B. Lewin in his book "*Genes*", "Most mutations that change the amino acid sequence are deleterious and will be eliminated by natural selection. Few mutations will be advantageous, but those that are may spread through the population, eventually replacing the former sequence. When a new variant replaces the previous version of the gene, it is said to become fixed in the population.... Sequence diversion is the basis for the evolutionary clock."

The mutational theory of evolution has the virtue of simplicity, and many (or most) evolutionists subscribe to it. Unfortunately, matters are not nearly so simple. As already mentioned, genetic variability can arise from many mechanisms other than outright mutation, e.g. genetic drift, sexual selection, incorporation of viral and bacterial genomes, symbiosis, and cultural practices. Mutation might, in fact, be only a minor contributor to evolution, particularly the evolution of intelligence. There is another complication that has been the source of confusion in the past: The causal relationship between genes and traits is not simple and straightforward. Beliefs to the contrary have led to published statements such as that by cell biologist Harvey F. Lodish, "Eventually the DNA base sequence will be expanded to cover genes important for speech and musical ability; the mother will be able to hear the embryo — as an adult — speak or sing." Claims of this sort ignore the fact that genes can be turned on and off, that identical sets of genes do not inevitably produce identical *phenotypes* (i.e. physically identical tissues or organisms), and that the development of most human traits depends upon the interaction of a complex gene network with their products (i.e. the proteins) and the environment. Almost 1000 genes affect human odor detection...but this does not seem to discourage outstanding scientists from predicting that we will be able to decode singing ability in embryos from DNA sequences.

Since evolution and mutation are interrelated in classical neo-Darwinism, it is useful to inquire into the frequency of mutations. Mutation rates have been estimated from so-called "divergence data" as illustrated by the following two examples: (a) All mammals are believed to have evolved from common ancestors about 85 million years ago. Today mammals show about a 10% divergence among their corresponding globin proteins (a special family of proteins). In other words, analogous mammalian globin proteins differ from each other at about 10% of their 100–200 amino acid sites. This corresponds to a replacement divergence of 0.12% per million years. Stated in another way, it took about 6–8 million years to alter one amino acid. (b) Two types of globins, α and β, are thought to have diverged from a single protein about 500 million years ago. The two proteins now differ from each other within a given species by about 50%, giving once again a protein modification rate of 0.1% per million years. The preceding numbers undoubtedly underestimate the overall incidence of mutations because (a) deleterious mutations have been removed by natural selection, and (b) organisms now have exquisite editing mechanisms for repairing errors that happen to occur in the DNA sequences.

E.T. Morch's classic work on mutational frequencies was based on the appearance of a dominant gene that causes achondroplasia (dwarfism). When parents, neither of whom are achondroplastic dwarfs and therefore lack the gene, have a child with the trait, the child must be in possession of a newly mutated gene. Frequencies of the dwarfism in the population indicate that about 4 out of 100 000 sex cells in normal persons contain the mutant gene. In another study of genetic change among clans of New Guineans since they arrived on their island 40 000 years ago, it was shown that between 2 and 4 base changes per 100 DNA bases have occurred per million years.

Little more will be said of mutation rates because of the general complexity and uncertainties of the topic. For example, there are so-called "hotspots", namely DNA sites where the mutational frequency is very much enhanced. As will be discussed briefly later, several recent papers claim that mutations need not be "random" but, instead, can be "directed" (i.e. occur at much higher rates when the newly acquired trait is advantageous). And there is no simple relationship between mutation and the degree of change that it imparts to the organism. Thus, a single-base ("point") mutation in a gene for an enzyme might have only a minor effect if the mutation converts an amino acid to another one of similar structure or if the change in the protein occurs far away from the small section of the enzyme where the catalytic reaction takes place. (The chapter on Molecular Evolution discusses this point). On the other hand, a gene mutation that affects the timing of an embryological process could have a dramatic effect upon the organism, an effect that (judging from the thousands of *Drosophila* fruit fly mutations that have been observed) would likely be either trivial or pathological. Compounding all this complexity is the fact that genes are often linked in cooperative units ("*operons*") and do not, therefore, act as independent entities; mutations in one component gene could, therefore, affect the expression of others.

In summary, simple "point" mutations have definitely played a role in evolution. Perhaps this role is less than is presently touted in the textbooks because there are many sources of genetic change. In any event, the relationship between mutation and evolution cannot be easily quantified owing to complexities and uncertainties. Should there be the slightest doubt in this regard, consider the following questions: What would have been the consequences on evolution if gene mutation had occurred at twice its actual rate (whatever that was)? Would there be species on earth now unknown to us? Would dinosaurs have been able to keep pace with environmental changes and thus still exist? Would humans be even more intelligent? Was the rate of gene mutation even uniform over the ages? No one can answer convincingly even such basic questions.

It is tempting to assume that, indeed, a doubling of the mutation rate would increase (double?) the rate at which species evolve. Yet this is not a foregone conclusion. Increasing the mutation rate in the fruit fly by 15 000 percent via exposure to X-rays (equivalent, at least in frequency, to millions of years of "normal" mutation) has never produced anything other than another, often freakish, fruit fly. No new species has ever been created artificially no matter what the mutation rate.

Consider now a prehistoric species which, being at equilibrium with its contemporary environment, is stable with regard to its physical appearance and other characteristics (its "phenotype"). Mutational events continue to occur, but the only mutations that "stick" (i.e. remain in the DNA) are the ones that are not physically expressed. All mutations that manage to become expressed perturb the happy status of the organism and become eliminated by natural selection. Now non-expressed mutations are common. They can appear as "neutral substitutions" in a protein that do not affect the activity of the protein. Or they can appear as "pseudogenes" defined as inactive but stable components of the genome. Thus, the organism, while at "rest", can build up a large storehouse of genetic variation that fails outwardly to alter the organism. Suddenly, there is a change in environment (an ice age or a new predator, for example), and the organism, taking advantage of its genetic variability (perhaps by "turning on" its silent genes), accommodates to the change in an evolutionary flurry. In such a scenario, evolution has more to do with the rate of environmental change than with the rate of mutation available to accommodate the change. This important mechanism has been given the name "*neutral evolution*".

Many neo-Darwinists now believe that evolution has been propelled "one gene at a time" via random "point mutations". Numerous difficulties, however, beset random point mutations as the sole source of evolution. For example, as we have just seen, the mechanism is excruciatingly slow and inefficient; beneficial replacements of one amino acid for another are rarities. Of course, this problem can be explained away by the millions or billions of years available to evolution. But slow mutational rates over vast time periods fail to accommodate many known cases of rapid evolution such as parasites that quickly modify themselves to combat changing immune systems in their hosts. One senses, therefore, that major macroevolutionary events (such as the development of human intelligence) demand something more than simple and inefficient amino acid substitution directed by point mutations. One yearns for a different kind of mutation, one that brings about larger genetic changes in a shorter time period. This would both speed up evolution and provide more substantial phenotypical modification than seems possible with painfully slow, random, and inefficient alterations of proteins.

Fortunately, there is in fact another mutation mechanism: Reorganization of genetic material. As I will show, genetic reorganization provides a treasure trove of variability. Specific examples in Nature will be given in the next chapter, but for the moment I will simply lay the groundwork necessary to appreciate this second source of variability. I will do so in a concise and self-contained format that requires no special prior knowledge of genetics. As is true here and throughout science, the key concepts are usually simple and comprehensible. And grasping them will permit an understanding of some remarkable natural phenomena. Let me express the hope that those who decide now to continue no further will, nonetheless, be pleased with the insights into the wonders of Nature that this book has described in the previous pages.

I begin with the definition of a "*transposon*": A transposon (or transposable element) is a DNA sequence that is able to insert itself at a new location in the genome. In other words, the gene is mobile. There need not be any relationship between the base sequence of the transposon and the region into which it inserts.

The simplest type of transposon is called an "*insertion sequence*" or IS. Listed below are some key features of the IS.

a) Roughly 1.5% of the genome of *E. coli*, a common bacterium, consists of IS elements.
b) IS elements are identified by the presence of "inverted terminal repeats" in which both ends of the mobile gene are flanked by the same sequence (commonly 9 bases) in reverse order (where numbers 1 through 9 below represent a particular sequence of nine bases such as AAGCGCAGC):

IS

DNA — 123456789-gene-987654321 — DNA

c) Most IS elements insert at target sites randomly, although some show varying degrees of preference for certain "hotspots". Insertion rates are comparable to spontaneous mutation rates. Displacement of an IS element thus occurs only infrequently. When an IS element inserts in the middle of a functioning gene, it can do a great deal of mutational damage.

d) About 3% of the *Drosophila* fruit fly genome consists of transposons at various sites. Reshuffling of a transposon occurs at a rate of once in every 10^4 to 10^7 cells. There are probably mechanisms that limit the frequency of transpositions in higher organisms because unrestricted shifting of genes could cause genetic havoc.
e) The frequency of transposition of a gene declines with the distance between the IS termini.
f) Simple gene relocation occurs by a "cut and paste" mechanism in which the termini of the transposon are cleaved, the target site in the DNA is nicked, and the transposon is inserted into to the nick (much like a short piece of film is spliced into a movie reel).

Mobile genetic elements ("*jumping genes*") were first discovered by Barbara McClintock in the early 1950s when she was examining the variegated pattern of Indian corn. Her proposal that genetic elements can move within the corn genome was ignored for 20 years by geneticists unable to depart from their ingrained prejudices that genes are fixes at a permanent location within the DNA. Eventually and belatedly, McClintock won the Nobel Prize for her research.

While on the subject of mobile genetic elements, mention should be made of so-called *exons* and *introns*. It turns out that most genes in higher animals are interrupted by "useless" sequences of bases. Thus, DNA segments of a gene that ultimately get coded into a protein (exons) are separated by DNA segments that never get coded into the protein (introns). The introns are never coded into proteins because their sequences get excised just prior to formation of the proteins from the genes.

The presence of introns greatly lengthens the gene. For example, the chicken ovalbumin gene has 7700 base pairs including seven exons separated by seven introns of varying sizes. After the introns are excised, the remaining useful genetic material, the sequence ultimately made into a protein, possesses only 1872 bases.

Why are introns present? No one knows. It is known that introns have no coding function. Moreover, intron sequences in corresponding genes among various species vary substantially, whereas exon sequences tend to be much more closely "conserved". The fact that higher animals, but not bacteria, have introns suggests that either (a) introns were inserted into the DNA of higher animals in the course of evolution or (b) introns were originally present in bacteria and then lost.

One final aspect of genetics should be discussed in concluding this short technical portion of the book. Its importance will be seen in the final chapter of this book where I propose an alternative mechanism for gene variation. Gene expression is encompassed by an intricate, multifaceted set of controls. Only a couple of controlling factors will be mentioned here. One of the most interesting of these is the *regulator gene*. A regulator gene codes for a protein involved in modifying the expression of other genes. The protein functions by binding to a specific site on the DNA and, by this means, either "*turning a gene on*" or "*turning a gene off*". In humans and other higher animals, the most common mode of regulation is of the "on" variety. Here, in brief overview, is how it works: A gene complex consists, in sequence, of (a) a regulator gene; (b) a promoter region; and (c) the "target gene" which codes for the actual protein (see the highly schematic diagram below). The promoter section of the DNA serves as a starting point for coding the adjacent target gene into a protein. In the resting state, the target gene is inactive (i.e. not coded into a protein). But the regulatory protein, created by the regulator gene, will recognize a portion of the promoter segment, bind to it, and thus initiate the coding of the target gene into the corresponding protein.

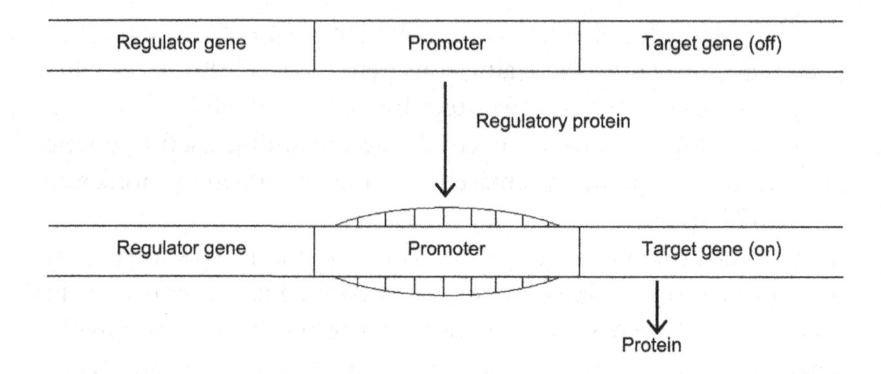

There also exist *enhancers* which, like the promoter region, assist in initiating the coding of the target gene, but they differ from promoters in that they are located at considerable distance from the starting point of the coding process. Distant repressors, called *silencers*, are also present. In addition to all this, cytoplasmic proteins (*signal transducers*) and steroid complexes can enter the nucleus from the cytoplasm outside the nucleus, bind to DNA, and induce or repress specific genes. The whole business is very complicated with the numerous transcription factors getting into the act. The bottom line, however, is simple enough: The expression of a gene (i.e. its directing the synthesis of a protein) may be actively accelerated or repressed through the effects of substances (including proteins) that bind to the DNA.

The "turning on" and "turning off" of genes explains a question which, I recall, I wondered about as a student: Why are a given organism's cells so different from tissue-to-tissue when all the cells have identical genomes (i.e. the identical DNA sequences)? The basic answer is now clear, namely that cells of different tissues express different sets of genes among all those that are available in the genome. For example, about 12 000 genes are expressed in common by both chicken liver cells and chicken oviduct cells. But, in addition, about 5000 genes are expressed exclusively in liver cells, while 3000 other genes are expressed exclusively in the oviduct cells. Thus, the masking and unmasking of genes are vitally important processes in embryological development.

Embryological development, in turn, is key to an understanding of evolution.

Recent work by geneticists M. V. Rockman and M. W. Hahn has shown an interesting role of regulatory DNA in brain chemistry. Prodynorphin is a protein that serves as a building block for chemical messengers in the brain known as endorphins. Endorphins have, in turn, been implicated in learning, memory, emotional bonding, etc. Now the identical prodynorphin gene is found in all primates. But humans possess two to four copies of DNA sequences that regulate prodynorphin production (i.e. four genes to make the same brain protein), whereas non-human primates have only one copy. Evolution has given humans an advantage, at least in this particular aspect of brain chemistry.

Gene expression and rearrangement must have been critical to evolution as well. The great majority of genes regulate DNA expression rather than lead directly to enzymes and other useful proteins. Among the silent (or "*cryptic*") DNA sequences, which comprise as much as 98% of the human genome, one finds a large proportion of transposons and non-mobile remnants of transposons. Being unexpressed, the sequences accumulate mutations at a greater rate than the expressed sequences which are under selective pressure and subject to constant repair. Potentially, this unexpressed DNA represents a great source of variability if, at some point, the sequences happen to become expressed.

The role of gene expression in humans can be seen strikingly in human embryos with their gill slits. Apparently, we still retain ancestral "fish genes" that manifest themselves for only a brief period before becoming suppressed later in the embryo's life.

Such is the genetics needed for the remainder of the book including my discussion of the evolution of intelligence. I will begin in the next chapter with a discussion of five common examples of gene variability and regulation. Gene variability and regulation are important topics because human intelligence evolved rapidly from apish beginnings with only a small overall alteration in our genetic code judging from the similarity (i.e. 99% overlap) of human and chimpanzee DNA. The most likely "explanation" for the dramatic improvement of human brain function is that there have been critical changes in genes that exercise control of our fetal development, i.e. genes that regulate the expression of other genes. Although understanding of fetal development, and the genes that regulate it, is severely limited, we do know something about rapid gene variability in snapdragons, sleeping sickness, the immune system, etc. These amazing and instructive products of Nature are briefly discussed in the next chapter. Gene variability and regulation will then become the basis for a new perspective on evolution in **Section 4**.

Chapter 13

GENE VARIABILITY, EXAMPLES

Part 1: Snapdragons

The following information on snapdragon genetics was taken from C. Willis' wonderful book entitled "*The Wisdom of Genes*". It turns out that the most common variety of snapdragon is dark red owing to the production of a red pigment in the flower. There also exists a largely ivory-colored mutant in which a gene involved in pigment synthesis has been altered. Actually, the ivory-colored mutant flower is covered with small red spots. These red spots are attributable to hundreds of scattered cells where the ability to synthesize the red pigment has reasserted itself via "*back-mutations*". But random point mutations would be orders of magnitude too rare to explain how an inactive gene could revert back to a pigment-producing gene. What is going on?

A snapdragon happens to have several copies of a transposon, named Tam3, inserted at various places within its genome. The ivory color results from a translocation of Tam3, in which it nestles its way in front of the gene coding for an enzyme involved in the synthesis of red pigment. This impairs the gene's function and thus the production of red pigment. But unlike most mutations, and this is key, the mobile Tam3 is able to jump away from its new location at a high rate. In so doing, the ability of the snapdragon cell to synthesize the red pigment is restored. In other words, each red spot of the ivory flower represents a cell in which Tam3 has departed from its inhibitory site adjacent to the gene for red pigment.

The red spots on the flower petals constitute *somatic mutations* which means that the mutations occurred in the plant somewhere other than in a germ cell of the plant; hence they are not inheritable. But if Tam3 in a germ cell of an ivory flower jumps away from the red pigment gene, then this back-mutation will be passed on to the next generation whose flowers will be totally red. If the removal of Tam3 from an ivory sex cell is not "clean" (i.e. if bits of the transposon are left behind), then the new generation will have a color intermediate between red and ivory. Thus, imperfect removal of Tam3 can generate mutants with a range of colors depending upon how much of the color-inhibiting transposon remains adjacent to the critical red-color gene. And all of this genetic variability is available orders of magnitude faster than the conventional mutation rate.

Ivory mutants grown outdoors have many more red spots than those grown inside a greenhouse where the temperature is higher. Low temperature clearly favors back mutations both in the petals and in the germ cells within the seeds. Suppose for the sake of argument (and there is no evidence for this) that red spots help absorb heat and thus protect the snapdragon from low temperature. We now have a fully functioning neo-Lamarckian mechanism: A normally non-mutational environmental perturbation (i.e. low temperature) induces a heritable change (red color) that protects the plant from that very same environmental perturbation. Is it possible, I will ask in the next section, that Lamarckian concepts play a role in evolution?

Part 2: Sleeping Sickness

Sleeping sickness is transmitted via a tsetse fly bite that infects humans with a unicellular parasite called a trypanosome. The disease, which may take two or three years to develop, begins with lassitude and culminates, as the parasite bores its way into the nervous system, in coma and death. Large areas of equatorial Africa are rendered uninhabitable because of the tsetse fly.

The trypanosome covers itself with a coating of a "surface glycoprotein" (i.e. a sugar-protein complex). It is by means of this glycoprotein that the host's immune system recognizes the trypanosome as a foreign intruder that needs to be destroyed. The trypanosome, however, has an ingenious mechanism for combating the immune system of its host: After multiplying in the bloodstream for about a week, the parasite synthesizes a new and "immunologically distinct" glycoprotein shell. That is to say, the host's antibodies that reacted against the original glycoprotein do not react against the new one. And the ruse continues. Every one or two weeks, a new glycoprotein is synthesized, thus presenting the host's immune system with an ever-varying challenge to recognize the parasite. More than 100 different glycoproteins are available in the trypanosome's arsenal. Since the host's immune system always lags behind the ever-changing glycoprotein on the trypsanosome's surface, the parasite can perpetuate itself definitely with little interference.

The interesting question, of course, is how exactly does the trypanosome genome manage to code sequentially and rapidly for a large number of different glycoproteins. It is known in this regard that each individual trypanosome carries the entire 100-gene glycoprotein repertoire within its genome. Consequently, a newly formed glycoprotein is not the product of a newly formed gene but, instead, it is the product of a preexisting gene that has been suddenly aroused from an inactive state. Simultaneously, the gene for the prior glycoprotein coating is silenced.

Production of one and only one glycoprotein is associated with a so-called expression site within the trypanosome DNA. When an inactive glycoprotein gene is shifted to an expression site, it becomes active. Thus, a trypanosome replaces a glycoprotein by (a) duplicating one of its 100-glycoprotein genes and (b) using the copied gene to displace the glycoprotein gene currently occupying the expression site. The order in which the 100 glycoprotein genes express themselves one-by-one is erratic and unpredictable; were it otherwise humans might have had an easier time evolving a defense mechanism.

Special note should be taken of the fact that trypanosome survival depends on genes turning on and off in a matter of days.

This rapid change in gene expression is important in that it shows that evolution need not be a tortuous, excruciatingly slow one-amino-acid-at-a-time-over-vast-time-periods affair as advanced by many neo-Darwinists. Genetic change is much more creative and efficient than this, a notion that I will apply to the evolution of human intelligence.

Part 3: Immune Diversity

No component of Nature is more wondrous than the immune system. Just think of what a person's immune system accomplishes: It genetically codes for and produces proteins ("*antibodies*") that can combine with, and thereby inactivate, literally millions of unwanted viruses, proteins, organic compounds, etc. (called "*antigens*"). These antibodies form readily even though a person has had no prior contact with the antigen. Indeed, antibodies are created in response to antigens that have never before been encountered by the human species during the entire course of our evolution! We know this is true because antibodies form when an individual is exposed to compounds that have been synthesized in the laboratory for the first time ever and are, therefore, totally unknown to Nature in the past.

Now an antibody inactivates a virus or other antigen by "grabbing" it in a tight, highly specific complex. A "lock and key" have been mentioned for want of a better analogy. Thus, an antibody for one virus will not in general serve as an antibody for another type of virus, just as a lock is specific for only one key. How in the world can the genes of the immune system code for proteins that will specifically combine with any antigen that might happen to assault the system? Where does this diversity come from? In answering these questions, I will simplify the situation to its bare essentials. This is all that is necessary in the context of the main thrust here, namely the possibility that the mechanism of immune diversity might be germane to evolutionary diversity.

It is usually assumed that humans have the ability to manufacture 10^6 to 10^8 different antibodies. Supplying this number of protein molecules using 10^6 to 10^8 different genes would require an absurdly large amount of DNA (seeing as humans actually have only about 20 000 genes). Now it turns out that each antibody protein consists of four subunits: Two identical light chains ("L") plus two identical heavy chains ("H"). If any L can connect to any H, then the necessary number of antibodies could be achieved with 10^3 to 10^4 different L's and 10^3 to 10^4 different H's, i.e. $(10^3 - 10^4)^2 = 10^6 - 10^8$. But even 10^3 to 10^4 genes for each L and each H would be too much of a genetic burden. Diversity evolved in another, more ingenious way as we shall see.

An antibody's two L chains and two H chains each possess two main sections: A "*variable region*" (V) and a "*constant region*" (C). It is the total of four V regions that surround and bind the antigen and thus constitute the recognition site of the antibody. Since the four V regions is variable, so is the specificity of the antibody.

Actually, I have oversimplified things a bit. With L chains, there is a short J region (J for joining) between V and C to give a V-J-C configuration. And the heavy chains have an additional D region (D for diversity) and can be represented as V-D-J-C.

To summarize:

$$L = V + J + C$$
$$H = V + D + J + C$$

This all seems complicated, but things will clarify momentarily.

And now to some approximate numbers. In germ cell DNA, and in every other cell including the B lymphocytes that manufacture antibodies, one finds the following gene cluster containing all the information needed to assemble the L-chain: An intron, 300 V genes, 5 J genes, an intron, and a C gene. (Recall that introns are useless pieces of DNA that ultimately get eliminated). The corresponding cluster for the H-chain consists of an intron, 300 V genes, 10 D genes, 5 J genes, an intron, and a C gene. The situation is thus summarized below:

L-chain intron — 300 V's — 5 J's — intron — C
H-chain intron — 300 V's — 10 D's — 5 J's — intron — C

Now I pointed out that all antibodies are composed of two L-chains and two H-chains. And all the 300 V genes differ from each other as do the members of the J sets and the members of the D set.

Finally, we arrive at the secret of immunodiversity. Each B lymphocyte (the producer of antibodies) selects one of the V genes and one of the J genes and splices them together, along with the constant C gene, into a new gene that codes for the L-chain. For example, a particular lymphocyte cell might select the 52nd V (from a list of 300) and the third J (from a list of 5) to form a V_{52}-J_3-C gene. Another lymphocyte cell might select the 38th V and the second J to form a V_{38}-J_2-C gene that codes for an entirely different L-chain. Similarly, a lymphocyte might splice together V_{10}-D_6-J_2-C or V_{235}-D_9-J_1-C to fabricate an H-chain gene (all components of which are taken from the original inherited gene "cluster"). In summary, the enormous variety of antibodies is the result of "cutting and pasting" from a standard cluster of genes, each lymphocyte cell doing this with a different combination of genes.

The joining of one out of 300 V's with one out of 5 J's can generate $300 \times 5 = 1500$ different L-chains. Actually, this number is low (perhaps by a factor of 5) because there turns out to be coding variations at the V-J juncture which also contribute to the overall diversity. This raises to $1500 \times 5 = 7500$ the number of possible L-chains. Similar arithmetic (and assuming now that the coding variation at the V-D and D-J junctures is worth another 100) gives $300 \times 10 \times 5 \times 100 = 1.5 \times 10^6$ different H-chains (from V, D, J, and the 100 "fudge factor", respectively). Since an antibody is made of randomly assembled L- and H-chains, the possibility exists of making $7500 \times 1.5 \times 10^6 = 11$ billion different antibody combinations. It is beginning to be clear why antibodies can interact with such a vast array of different foreign objects.

The immune diversity story does not stop here, however. Antibodies are subject to even more variations through the process of rapid "*somatic mutations*". Somehow, specific parts of the V gene can "hyper-mutate" (i.e. mutate at rates a million-fold higher than the normal rate of spontaneous mutation). This increases — by orders of magnitude — the number of possible antibodies beyond the 11 billion number estimated from the gene recombination mechanism alone. More than enough immune diversity is thus available to counter most of any viral or other antigenic assault (except when the virus or parasite mutates too rapidly for the immune system to keep up).

If any particular antibody is produced by only a single B lymphocyte cell, then how can there be enough antibody to stave off a major dose of antigen as might arise in an infection? The answer is that a positive antigen/antibody interaction triggers a "*clonal expression*" of the lone B lymphocyte cell producing that particular antibody. In other words, the relevant B lymphocyte cell, sensing the foreign virus or whatever, proliferates without further mutation, thereby providing (hopefully) sufficient amounts of the correct antibody to deal with the infection. Of course, when the virus or parasite mutates too rapidly for the immune system (that is to say, the clonal expression of suitable B lymphocytes cannot keep up), there will be a particularly serious health problem.

The immune system teaches some important lessons. New antibody genes are assembled from widely scattered pieces of the genome and then further modified by facilitated mutations. Thus, new genes, and the clones that own them, are available quickly in response to environmental factors (e.g. a foreign bacterium). Granted, the new genes are created in the genomes of B lymphocytes and not in the germ cells. That is why disease resistance is not inherited but, instead, acquired individually. Yet there is no reason why, at least in principle, a mechanism operative in the somatic cells could not also function in the germ cells. If this happened, then all the elements of a true neo-Lamarckian process would be in hand: (a) a potential for genetic variability in germ cells based on rapid non-mutational mechanisms; (b) modification of the DNA of germ cell precursors in response to an environmental perturbation or to biochemical side-effects of such a perturbation; (c) a proliferation of the modified germ cell precursors into a population of fully formed germ cells; (d) the passing on of the new genes to future generations; (d) a possible gradual diminution of the new trait with time (as happens with antibodies) in generations not exposed to the environmental perturbation that initiated the whole cascade of events in the first place.

In immuno-diversity, one particular gene combination (producing one particular antibody) is replicated in large numbers in the presence of an antigen. In the neo-Lamarckian counterpart, one particular gene combination, stimulated by an environmental perturbation or its biochemical by-products, is produced in large numbers among the germ cells. The mutation is, therefore, extended to the next generation. The subject will be covered in great detail in **Section 4**.

There is no hard evidence, that I know of, proving the occurrence of an immune-style mechanism within germ cell tissues. Lack of evidence, however, does not constitute disproof. The only requirement at this point is an open mind willing to accept, within limits, the possibility of undiscovered and unconventional biological mechanisms.

Part 4: Globins

We have seen that snapdragons can back-mutate quickly; that trypanosomes can turn genes on and off in a matter of days; that the immune system can assemble genes from scattered fragments and proliferate useful combinations within hours. Globins constitute yet another remarkable example of genes being far less static than is generally invoked in the neo-Darwinian model.

Mammalian hemoglobin, a protein that gives blood its red color, binds inhaled oxygen and then releases the oxygen in tissues that need it for metabolism. Adult human hemoglobin is a "tetrameric" protein, i.e. it consists of four protein chains comprised of two identical α chains and two identical β chains. The protein can, therefore, be symbolized as $\alpha_2\beta_2$. Now one might think that the genetics of the situation would be simple (a gene for α and a gene for β), but in fact the situation is more complicated than this. There is a gene "cluster" or "family" coding for several different types of α protein chains; the same is true for several types of β chains. The two families of genes are spatially separated from each other in the genome.

The two gene clusters are given schematically below. The symbol ψ represents so-called "*pseudogenes*" that are stable but inactive components of the genome derived from mutations of once-active ancestral genes. Additional non-functional (useless) genetic material of varying length (symbolized by "..." in the scheme) serves to separate the functional (useful) genes (ζ, α, ε, δ, and β), each of which codes for a protein belonging to the α-type or β-type gene cluster. For example, ζ is an α-type gene, and ε is a β-type gene.

α-type cluster: $\ldots\zeta\ldots\psi\ldots\psi\ldots\psi\ldots\alpha\ldots\alpha\ldots\psi\ldots$
β-type cluster: $\ldots\varepsilon\ldots\gamma\ldots\gamma\ldots\psi\ldots\delta\ldots\beta\ldots$

The point of all this is that humans actually produce different hemoglobins in three developmental stages: embryonic, fetal, and adult. Hemoglobin $\zeta_2\varepsilon_2$ appears in the embryo of less than 8 weeks. After 8 to 12 weeks, the ζ and ε genes are turned off, while the α and the γ genes are switched on. Thus, the 3–9 month fetus possesses $\alpha_2\gamma_2$ hemoglobin. Finally, adult $\alpha_2\beta_2$ hemoglobin is produced at birth by turning the γ gene off and turning the β gene on.

Clearly, genes are susceptible to carefully regulated, time-dependent switching mechanisms. Once again, control circuits for gene induction and repression are seen to be readily available in human genetics. And, importantly, the changes in gene expression are heritable; despite innumerable cell divisions as an adult, the ζ, ε, and γ genes for fetal hemoglobin proteins are never again expressed. For all practical purposes, these genes have been repressed...*not mutated*...out of existence.

Part 5: Heat Stress

Humans begin their life as a single cell (a fertilized ovum) and end up with a host of differentiated cells comprising the liver, heart, muscle, and other tissues. Each differentiated cell expresses only a small fraction of the total number of genes potentially available to it. Exactly which set of genes is expressed in a differentiated cell depends upon the particular cell type. The selectivity of gene expression among an organism's cells is absolutely astounding. Thus, cells from two different tissues, although possessing identical genomes, can differ in their synthesis levels of a particular protein by a factor as large as 10^9. For example, reticulocytes (immature red blood cells) synthesize large amounts of hemoglobin but no insulin. Pancreatic cells produce, in contrast, large amounts of insulin but no hemoglobin. A tissue-specific protein expressed by the testes and nowhere else (called β-tubulin) is required for assembling a highly specialized component of sperm tails.

Obviously, a host of genes, whatever the cell type, must be "*silenced*" via some negative regulatory mechanism. Two mechanisms, "*passive control*" and "*active control*", can be envisioned. These will be discussed in turn.

Passive control shuts down unneeded genes for the life of the organism (as we saw happen with fetal hemoglobin). Once a gene is silenced during a developmental stage of the organism, no more is heard from the gene again. Permanent, irreversible inactivation of this sort has an appealing simplicity and stability (and one wishes that the mechanism were understood in greater detail).

Active control implies an on-going regulation of gene expression. Thus, expression and non-expression are reversible, reflecting the cell's content of regulatory proteins at any given time. Genes under active control are said to exhibit "*plasticity*".

We have only a rudimentary understanding of how genes relate to adult organs such as the brain. Accumulating evidence shows, however, that active control does influence the expression of genes involved in cell differentiation. Gene plasticity seems, therefore, to be a widespread phenomenon. Several examples below further illustrate the ability of genes to switch on and off.

a) Heat-sensitive *E. coli* bacteria can change their gene activity when exposed to an increase in temperature. Thus, a temperature increase causes *E. coli* to turn off genes that direct the synthesis of a particular set of proteins and, simultaneously, to turn on "heat shock genes" that produce 17 new proteins. A key role is played by a regulator gene and its protein product (σ32) that trigger the heat-stress response. The regulator gene, in turn, is stimulated into action by the presence of proteins that have been unfolded (denatured) by the temperature increase.

b) A striking example of gene plasticity concerns the central nervous system of *Drosophila* fruit flies. When an adult female is exposed to a shift in temperature, the expression of its "tra-3 gene" is disrupted, causing the female to engage in a complex courtship ritual typical of the adult male.

c) Humans are either male or female according to whether they possess XY or XX chromosomes, respectively. Alligators, on the other hand, have no X and Y chromosomes. Their sex is determined by the temperature of the nest in which the eggs are incubated. Cooler nests produce females, while warmer nest produce males; intermediate temperatures can produce both sexes. Apparently, temperature permanently affects the expression of sex-determining genes at some early stage of development.

d) Temperature, of course, need not be the only environmental factor affecting gene expression. Specialized root cells from the carrot will, when bathed in coconut milk, begin to develop into carrot plants with normal roots, stems, flowers, and seeds. The genes necessary to create a mature plant are all turned on by coconut milk! In a remarkable process called "*metamorphosis*", tadpole genes are switched off and frog genes are turned on. And a wound in the skin switches on the genetic machinery necessary to stimulate cell growth and wound healing. It does so with high precision; seldom does the healing of a minor wound lead to an abnormal depression or lump in the skin.

e) A final example of control elements at work is based on the fact that the frog *Xenopus* produces different sets of proteins in its cultured somatic (non-germ) cells and in its oocytes (egg cells). The two characteristic protein patterns are easily distinguished

> by a technique called "two-dimensional gel electrophoresis". Now when a nucleus from a cultured *Xenopus* somatic cell was injected into the oocyte of *Pleurodeles* (a salamander), the guest nucleus began directing the synthesis of proteins with the oocyte pattern, not the somatic cell pattern. Thus, the cytoplasmic environment of the salamander egg was able to inactivate the expression of one set of frog genes (i.e. the somatic cell genes) and to activate the expression of another set characteristic of the frog oocyte.

Evolution might have proceeded by gene rearrangements (not unlike those seen with the snapdragon and immune system) and by altered gene expression via the effect of control elements (not unlike those seen with globins and heat stress). This is hardly a new idea. Unfortunately, in most cases the relationship between gene expression and anatomical trait is unknown with vertebrates owing, in part, to the extreme complexity of multigene families and their control systems. Yet there is a certain "advantage" to this ignorance: It allows a degree of freedom when I speculate about the origins of the most puzzling trait of all: Human intelligence.

Chapter 14

DIRECTED MUTATIONS

It is a fundamental tenet of evolutionary biology that any particular mutational event is independent of its future value to the organism. Mutations, in other words, are random. Natural selection tends to increase the frequency of those rare mutations or genetic changes that happen accidentally to impart survival or reproductive advantage to the organism. This is the heart and soul of neo-Darwinism.

In the past decade, there have been claims that certain mutations in bacteria and yeast occur more often when the resulting phenotype is advantageous. Such mutations have been described as "*directed*" (or "Cairnsian" in reference to J. Cairns, a major proponent of directed mutations). The idea of directed mutations, antithetical as it is to the neo-Darwinian tradition, engendered considerable controversy that is still simmering. The directed mutation is discussed here because (right or wrong) the idea is a fascinating one and because it is amazing that, in these modern times, there can be strong disagreement among respected biologists over something as elementary as mutations.

Let me begin with the experiments of B. G. Hall with *E. coli* bacteria that are normally unable to metabolize a compound called "salicin". *E. coli* has all the genes to make salicin-metabolizing enzymes, but the genes cannot be accessed. In a word, the genes are silent or cryptic. Hall found two mutational events that are necessary to allow the bacterium to metabolize salicin: (a) a spontaneous mutation in a regulatory gene and (b) a removal of an IS element (see *Chapter 11*) from one of the genes coding for an enzyme. If both mutations occur, the *E. coli* can then metabolize salicin.

Spontaneous mutation in the regulatory gene occurs with a frequency of one in every 20×10^6 cells. Removal of the IS sequence is even more infrequent: One in every 5×10^9 cells. If mutations are truly random, then the double-mutation should be observed in one in every 10^{17} cells (i.e. the product of the two individual probabilities, and an extremely small number). On the average, one would need a cubic yard of *E. coli* bacteria to find a single double-mutation.

When normal *E. coli* were grown on a medium of salicin plus some other nutrients, the first few generations survived on the nutrients other than salicin. But, in only two weeks, two-thirds of the colonies could utilize salicin, a remarkable occurrence given the extreme unlikelihood of the required double-mutation. Hall found that the spontaneous mutation at the regulatory gene took place first at the normal rate. It was the excision of the IS that was abnormally fast. Apparently, the presence of salicin itself triggered the removal of the IS element from a gene critical to salicin metabolism. In other words, evolution via mutation had transpired according to the dictates of the environment, namely the availability of the salicin food.

Naturally, certain control experiments were carried out. No IS excision mutants formed in the absence of salicin. Mutation rates at other genes remained the same with and without the presence of salicin. And IS excision by itself without a prior regulatory gene mutation does not give bacteria capable of growing on salicin. Hall in *Genetics*, **120**, 887, 1988) expressed it this way: "Excision occurs only on media containing salicin, despite the fact that the excision itself confers no detectable selective advantage and serves only to create the potential for a secondary selectively advantageous mutation." Whether or not one agrees that Hall's data represent a clear-cut example of a "directed mutation", the brilliance of the experiments, and the willingness to depart from seemingly inviolable dogma, must be conceded. The press, of course, sensationalized the work as proving Lamarck's idea that acquired traits are inherited.

What has been the basis of the counter-arguments? J. E. Mittler and R. E. Lanski in *Nature*, **356**, 446, 1992 found that most excision mutants outgrew the non-mutated cells on media with salicin but not on media without salicin. This calls into question a Hall assumption that each cell with an excision mutant is the result of a separate and independent mutation. In other words, salicin may not direct an advantageous IS-removing mutation so much as stimulate the proliferation of those few that do form. The fast accumulation of advantageous double-mutants is noteworthy only if cells with the regulatory gene mutation cannot abnormally multiply under the selective conditions and, by this means, increase the probability of the second mutation.

Double-mutations are not the only source of support provided by the proponents of directed mutations. For example, J. Cairns found lactose-consuming mutants of *E. coli* (Lac+) accumulated over time when a strain of Lac– *E. coli*, unable to utilize lactose, were incubated on a medium containing lactose. They found no comparable increase in mutants at other gene locations, nor an accumulation of Lac+ mutants on medium without any lactose. Opponents of direct mutations claim that the experiments again fail to account adequately for population dynamics.

It would be pretentious for me to take a position in this debate. Suffice it to say that I do not see why, at least in principle, directed mutations should not be possible. After all, we have no clear understanding of the factors that control the relocation rates and sites of IS and other mobile DNA elements. Under such circumstances, one feels free to postulate whatever one likes including the possibility that the mobility of a regulatory element from a gene is promoted by the very compound whose metabolism is controlled by that gene. C. Willis would call this an example of "*evolutionary facilitation*", i.e. an evolutionary direction triggered by the very environment to which the organism must adapt.

Chapter 15

GENETICS AND INTELLIGENCE

Having discussed basic tenets of genetics, I can now confront the subject of intelligence and its genetic basis. Since the genetics of intelligence is understood only at the most rudimentary level, this chapter need not be a long one.

In one current model of intelligence, there exists a set of genes dictating intelligence. According to some theorists, these genes can have either positive or negative impact on intelligence. A person will have a high intelligence if he or she has many positive genes and only a few negative genes. There is, therefore, a sliding scale of intelligence (the well-known bell-shaped curve) depending upon the proportion of the two sets of genes a person happens to inherit.

Natural selection implies that genes important to survival tend toward constancy, whereas genes peripheral or irrelevant to survival tolerate considerable variation. Thus, since all cats have sharp claws and teeth, but variable color, one might surmise that claws and teeth were more critical to cat survival and evolution than color. In other words, strong selection for claws and teeth has exhausted their genetic variability. Now since intelligence is somewhat variable in any given population (although not, on average, among large populations), then one might conclude that evolution did not strongly select for intelligence, i.e. that intelligence is not, evolutionarily speaking, an important trait. But this flies in the face of reason. Two things are wrong with the preceding argument. First, intelligence variations within a given population are far less variable than, say, the color of a cat so that the gene-constancy criterion may not apply. Second, intelligence is such an ultra-complicated, multigene trait that again it is difficult to generalize anything from our modest degree of genetic variability.

The above difficulties notwithstanding, there have been well-funded, well-organized searches for intelligence genes. In one of them, directed by R. Plomin, DNA from high-IQ and low-IQ children was analyzed for somewhat arbitrarily selected 100 genes. In a word, the results were inconclusive. This is hardly surprising because the trait in question, IQ, has not even been rigorously defined. More importantly, an effect of a gene variation on intelligence does not prove a causal relationship between the two. Consider, for example, a mutation in new-born babies that impairs the metabolism of an amino acid (phenylalanine). As a result of the mutation, chemical intermediates build up in the brain, and mental retardation (a condition called phenylketonuria or PKU) results unless the diet is kept free of the amino acid. Although we have here a direct connection between intelligence level and a specific mutation, this does not mean that the mechanism of intelligence development is fundamentally related to this gene.

There is still another difficulty with gene-by-gene searches for intelligence. Genes for complex traits do not act independently of other genes and the environment. Evolutionary geneticist Richard Lewontin said it well:

> Genes in populations do not exist in random combinations with other genes ... The fitness of a single locus ripped from its interactive context is about as relevant to real problems of evolutionary genetics as the study of the psychology of individuals isolated from their social context is to an understanding of socio-political evolution. Context and interaction are of the essence.

Thus, searches for genes governing intelligence, political inclination, ambition, criminality, and a host of other human traits are doomed to disappointment.

Geneticists now realize that traits, especially complex traits such as intelligence, are affected by gene networks composed of tens or hundreds of genes and gene products. These genes and gene products interact with each other and, as a group in combination with environmental factors, control the development of the trait. To predict a change in a complex trait from the mutation of a single gene would be like predicting the change in the economy of a town from a change in spending habits of a single resident. To summarize: Gene networks, not single genes, should be considered the unit of evolutionary variation. In practice, this principle currently forces us to consider how a single chance event (i.e. a mutation) affects the development, regulation, and stability of multiple interactions all focusing on a complex trait in a series of gigantic feedback mechanisms. The challenge here is awesome; a satisfactory understanding will not come to us quickly.

Section 4

EVOLUTION OF INTELLIGENCE, AN EPIGENETIC MODEL

Chapter 16

INTRODUCTION

If this book were a novel, one would have to say that, at long last, it has finally reached its "*dénouement*" (defined by Webster as the final revelation which clarifies the outcome of a plot). Of course, this is not a novel, and there is no plot — only an extensive discussion of the principles of Darwinism, anthropology, and genetics that help place what is about to follow in some sort of context. It may seem strange that a major portion of this book is, in a sense, introductory, or at least preparatory, in nature. But a treatise that combines uncertainties of evolution with the mysteries of the mind requires a secure foundation. Only then can a newly proposed mode of evolution be accepted or rejected intelligently.

One thing must be made clear immediately. The proposed mechanism for the evolution of intelligence in this section is pure speculation. There is absolutely no evidence for the construct at the present time. This being the case, what is the purpose of the section? Why bother? The answer to this question is simple: I want to show that mechanisms distinct from mutation-based natural selection are at least *possible* and *reasonable*. Someone once said that he would rather have a bad hypothesis than no hypothesis. This is because a bad hypothesis at least stimulates thought and discussion, whereas no hypothesis is sterile and non-productive. Thus, even if my hypothesis is bad (and I do not think it is), I prefer it to an absence of alternatives to natural selection theory and its drawbacks. In summary, bear in mind that **Section 4** merely guides you through a world of conjecture. Fortunately, this should not come as too much of a shock because much of the material in the preceding sections of the book on natural selection has entailed conjecture.

Before presenting an alternative model, let me attempt a one-paragraph summary of this book as it has unfolded thus far. **Section 1** summarized the basic tenets of natural selection. It was argued that most of the common criticisms of Darwinism are either inconclusive (e.g. the gap problem) or outright false (e.g. the entropy issue). **Section 2** discussed many aspects of the human mind, particularly our linguistic, musical, and mathematical intelligence. A search into possible origins of this intelligence came away with one main conclusion in direct conflict with Darwinian theory: Our intellectual power far exceeds any past need of our ancestors who lived as hunter-gatherers for the bulk of their evolutionary history (and this, in no way, is meant to discredit the cleverness of our ancestors). For example, human brains develop a center devoted to written vowels although writing is only a few thousand years old. An ability to speak five languages is commonplace among humans. Our musical ability, such as operatic singing and playing four or five melodies simultaneously on the piano, defies any imaginable survival or reproductive advantage, as does the previously cited ability to mentally multiply two 13-digit numbers. To quote from *Chapter 8*: "Natural selection is faced with the problem of explaining how a *complex set of genes*, controlling an *expensive trait* with *no obvious benefit*, came into *permanent existence* in such a *short time period* within *every member* of a *small population* (that was *dispersed and geographically isolated* over the entire planet) who had a *low reproductive output* and a *low mutation rate*. Finally, **Section 3** extracted from the vast field of genetics a few common biological processes, such as antibody production, where a high degree of gene variability and expression (far exceeding the formation rate of point mutations) is of paramount importance.

Each of the three previous sections covers a different subject and stands by itself. It is possible, if one so chooses, to read a section of particular interest and ignore the remainder of the book.

Acquaintance with all three sections, however, should help the reader better appreciate these final pages describing new ideas on the evolution of human intelligence. The construct is based on "*epigenetics*", and I begin with a brief discussion of that topic. Relating evolution to epigenetic changes is not a new idea, but nowhere has anyone proposed a detailed epigenetics-based mechanism for the evolution of intelligence.

Chapter 17

EPIGENETICS

"*Epigenetics*" literally means "outside genetics". Somewhat more useful descriptions, expressed in different ways, are given below:

a) Epigenetics is the study of traits that are heritable but do not involve classical mutations at the DNA level.
b) There is a second inheritance system — an epigenetic system — in addition to the system based upon the precise DNA sequences.
c) An epigenetic change is a change in the properties of an organism that is inherited but that does not represent a disruption in the total genetic potential of the organism.
d) Enduring phenotypic modifications (i.e. heritable changes in the appearance or other characteristics of an organism) reflect changes caused not only by DNA mutations but also by non-genetic or "epigenetic" mechanisms.
e) There is a solid body of evidence demonstrating that enduring modifications can be induced in the properties of cells via the presence of environmental stimuli; these modifications continue to be expressed for many cell divisions after withdrawal of the inducing stimulus.

Epigenetic mechanisms were initially postulated to explain the changes in cellular characteristics ("*phenotype*") during embryological development. Thus, a fertilized human ovum divides again and again, ultimately differentiating into a variety of radically different, specialized cells (nerve, muscle, kidney, liver, blood, etc.). But heritable differences among the various cell types cannot reside in the DNA sequences per se because all the cells in a given individual have the identical DNA derived originally from the fertilized ovum. Some factor outside classical genetics, termed epigenetics, must be postulated.

Epigenetics can be considered the "memory" system by which cells with different characteristics, but with identical DNA ("*genotypes*"), transmit their characteristics to their descendants *even when the stimuli that originally induced these characteristics are no longer present.* Thus, skin cells divide to give skin cells, liver cells divide to give liver cells, muscle cells divide to give muscle cells (after the embryological factors that originally converted the cells into their respective tissue types are likely long gone). This all sounds neo-Lamarckian, and indeed it is (and it explains why I devoted a long chapter to Lamarck). Lamarck, ridiculed for two centuries, may ultimately end up with the last (celestial) laugh! But, before this happens, biologists must accept a new logic, a logic that differs radically from current ideas on heritable cellular adaptations.

It is often difficult to differentiate a genetically-based trait from an epigenetically-based trait, a point illustrated by the work in the 1940s and 1950s of Sir Cyril Hinshelwood. After Hinshelwood won the Nobel Prize in chemistry for his work on reaction kinetics, he began experimenting with bacteria. Bacteria were grown on sub-lethal doses of drugs. Most, but not all, the bacteria survived, and the survivors were then repeatedly transferred to fresh media containing the drugs. Hinshelwood observed that the bacteria gradually adapted to the drugs to an extent depending upon the number of serial passages to which the bacteria had been exposed. After sufficient number of passages, 100% of the bacteria survived the drugs. These resistant bacteria were then grown for several generations on drug-free media. When the bacteria were transferred to drug-containing media, they survived, indicating that the original resistance had been maintained during growth and multiplication in the drug-free media. Hinshelwood concluded that he was observing a heritable adaptive change or, in modern lingo, an epigenetic modification.

Unfortunately for Hinshelwood, however, his conclusions met with wholesale dismissal. Geneticists felt that his adaptations had arisen from spontaneous mutations rather than from heritable environmentally-induced effects. Undeterred, Hinshelwood amassed a host of additional examples where bacterial changes occurred in response to almost every conceivable kind of sub-lethal chemical and nutritional stress. Although it is difficult to outright eliminate contributions from mutational events, several features of Hinshelwood's data favor an adaptive epigenetic mechanism: (a) The doses were sub-lethal, so this is not a case (as with classical antibiotic resistance) where all the bacteria are killed off except for a few resistant mutants that continue to multiply. (b) The drugs and nutritional compounds are not known to be mutagenic. (c) Rather than an "all or none" effect characteristic of many mutations, there was an almost continuous increase in resistance as the number of passages on drug-containing media progressed. (d) Resistance was expressed more quickly throughout the population than would be expected from rare mutational events. (e) Unlike most mutational behavior, the bacteria eventually regained their original drug-sensitive phenotype after they had been grown for many generations on drug-free media. (f) Similar results were obtained with a large number of chemically unrelated drugs and artificial nutrients.

None of the above features smacks of random mutations. But the results are consistent with epigenetically-based metastable and heritable changes in cell phenotype caused by exposure to small doses of drugs and nutritional analogs. A drug appeared to induce a heritable non-mutational ability to inactivate that same drug.

A second interesting example will drive home the importance of epigentics: The existence of mutated genes that cause cancer, called "*oncogenes*", is firmly established. There are, however, animal tumors that are initiated in cells without an apparent change in the DNA. Consider the case of embryonal carcinoma in the mouse. When such tumor cells that had been injected into a normal blastocyst (an early embryological stage) were placed in a mouse uterus for further growth, no tumor appeared. Instead, the tumor cells participated in a normal development, and they grew into normal tissues. Clearly, the expression of this cancer phenotype depends upon the cellular environment. Since, under the appropriate conditions, the cancer cells function as normal cells, the cancer cells must possess a gene that can be easily turned off or on by epigenetic factors rather than by the DNA sequence per se.

Epigenetics is now playing a role in biotechnology. For example, genes inserted into a rat can be made to function only when the rat is given an antibiotic. Or a gene package that makes human growth hormone does so only in response to a commonly available drug. The point here is that, in the words of S. J. Gould, "The genome is fluid and mobile, constantly changing in quality and quantity, and replete with hierarchical systems of regulation and control."

Now that the concept of epigenetics has been established, it is worthwhile to return to those forgotten or ignored Lamarckian experiments described in *Chapter 4*. Is it possible that Pavlov's transmission of conditional reflexes was indeed inherited? That Schroeder's caterpillars did in fact learn to "side-roll" a leaf and pass on the trait to their descendants? That an inherited transferal of an acquired immunity was indeed observed by Steele and Gorczynski? That poor Kammerer really did create heritable nuptual pads on *Alytes* by prolonged exposure of the frogs to water? In view of our current (rudimentary) understanding of genetic variability and of epigenetic effects, these Lamarckian explanations seem more palatable.

There are at least four types of epigenetic inheritance systems (see E. Jablonka and M. J. Lamb, "*Evolution in Four Dimensions*" for further details): (a) self-sustaining loops; (b) RNA interference; (c) structural inheritance; and (c) chromatin marking. For the purposes of this book, only the first two types need to be mentioned. Self-sustaining loops in biology can be most simply exemplified in the scheme below. Gene A (xxxx) is seen to have an adjacent promoter region (-----) of the type discussed in *Chapter 12.*

Assume that some temporary environmentally-promoted molecule T binds to the promoter, thereby activating gene A and inducing it to produce several molecules of protein B.

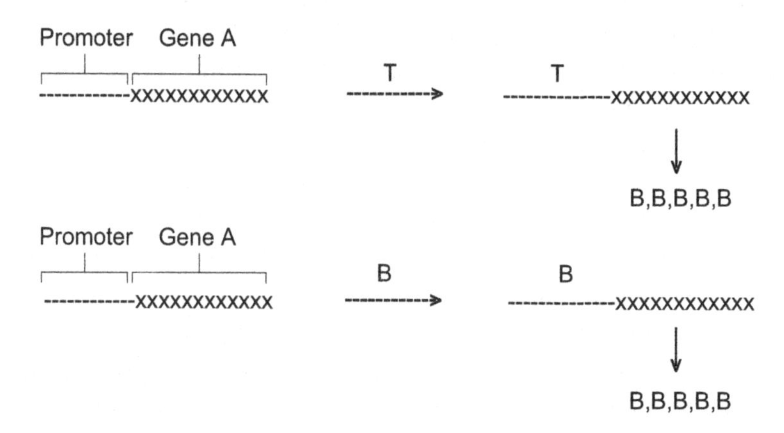

But B can likewise bind to the promoter, causing the production of more B from gene A and thus closing the "loop". In other words, once initiated by T, the production of B continues uninterrupted in a so-called "*feedback mechanism*". The mechanism is predicated upon B stimulating its own formation (in the field of chemistry this would be called "auto-catalysis"). Stated in another way, the product B of gene A can stimulate A (i.e. turn gene A on) to produce more B.

Now here comes the major problem with regard to invoking self-sustaining loops in epigenetic inheritance: The environmentally-based initiator T may no longer be present when the parent cell divides into the daughter cells. How does a daughter cell inherit the ability to produce B, like the parent cell can, when T is absent? The most obvious answer is that protein B is donated to the daughter cells when the parent cell divides. When a cell divides, it does more than donate nuclear DNA to each daughter cell; the parent cell also transfers the contents of its cytoplasm containing B, and in this manner the activation of gene A continues from generation to generation. As long as a supply of B from B-activated gene A continues unabated, succeeding generations of daughter cells will all have the B-producing trait. If, however, the supply of B gradually diminishes from generation to generation (for whatever reason), then gene A will be less and less activated with each new generation of cells, and eventually the cells will lose their ability to produce B unless the epigenetic factor T once again reappears and starts the whole process over again. This is a prime example of epigenetic inheritance in action.

Much the same mechanism can be applied to higher organisms. Suppose an epigenetic factor T stimulates the expression of gene A in an egg cell so that B is suddenly produced within this cell. After the egg cell is fertilized, it divides to initially form two cells on their way to creating a multicellular embryo. Each of the two cells has, roughly, half the amount of B present in the parent egg cell before the latter divided. But if this amount of B suffices to activate gene A in the daughter cells, each daughter cell can attain the level of B originally present in the parent egg cell via its own B-producing machinery. And the process continues as the cells divide further to form, ultimately, the adult. Thus, the ability to produce B, originally acquired after a temporary exposure of the egg cell to T, can be passed on to every cell in the organism by a self-sustaining loop. Importantly, the "B-trait" was inherited without the persistent presence of T and without mutational modification of the DNA sequences.

There is recent evidence that the total population of a cellular protein need not in fact be divided in two when a cell divides. The bacterium *Rhodobacter sphaeroides* produces a cluster of proteins that help direct the movement of the bacteria toward desirable compounds. The proteins are localized near the center of the cell's cytoplasm. It has been shown that the protein cluster is duplicated just prior to cell division, after which the two clusters position themselves at the centers of the newly formed daughter cells. In other words, the original protein concentration is maintained in the daughter cells via protein duplication. This constitutes another mechanism in which epigenetic initiator T can induce protein B which, in turn, persists at a constant level in future generations even after T has vanished from the scene.

The preceding model is simplistic in the sense that, as mentioned, most complex traits reflect the activity of not a single gene but a whole network of genes and their products all interacting with each other. Since, as seen above, an epigenetic inheritance mechanism can be constructed from only a single loop, one can imagine the intricacy of epigenetic mechanisms possible when tens or hundreds of interlocking loops come into play.

RNA interference represents a second epigenetic mode in which specific genes are silenced in a heritable manner. Suffice it to mention here that RNA is a gene-product with a structure similar to DNA. The important point is that when for some reason an abnormal RNA appears on the scene, the cell gets rid of it via an enzyme that chops it up into small pieces. The small pieces apparently bind to the same abnormal RNA from which they were derived, inducing further destruction. Additional interesting properties of the RNA pieces are: (a) They can be amplified to increase the number of copies. (b) They can migrate in the body, moving from cell to cell over long distances even among cells of different types. (c) Importantly, they can cause the inheritable silencing of the gene that produced the abnormal RNA in the first place. Thus, the mechanism provides a "double whammy" in that the small RNA pieces not only catalyze the destruction of the abnormal RNA, they also prevent its very formation. It is believed that the RNA inference mechanism evolved (a) to destroy unwanted RNA received from viruses and (b) to silence viral genes that do manage to get inserted into the host's genome.

When an abnormal RNA, corresponding to a specific gene in *C. elegans* (a nematode worm), was injected into the worm's gut, the gene was silenced by the RNA interference process. This silencing was transmitted through several generations of the worm's offspring. There is, therefore, some kind of non-mutational transfer of information, originating in the gut, to the germ line. In short, information flows from body cells to reproductive cells. This concept, which has a distinct neo-Lamarckian touch to it, will play a role in the epigenetic model of intelligence coming up next.

Chapter 18

THE CRANIAL FEEDBACK MECHANISM

Darwinism cannot accommodate traits that are, on the one hand, complex, expensive, widespread, and persistent and, on the other hand, of little or no immediate utility. Human intelligence, which exceeds simple survival and reproductive needs during hunting-gathering, is such a trait. Of course, examples of traits that are seemingly "excessive" are no doubt commonplace in biology. A flower may produce more red pigment than is necessary to attract pollinators. Or perhaps a spider might ensnare as many prey using a simpler, energetically less costly web. But the human brain is a so-called "organ-of-perfection" that consumes a significant fraction of our energy needs to own and operate. An untold number of genes are necessary to control brain development and function. One cannot, therefore, simply dismiss our extraordinary intelligence as random evolutionary "overshoot".

What follows is an epigenetic model that attempts to explain the evolutionary development of intelligence. As stated previously, the model lacks hard evidence. Thus, I will never argue that the epigenetic mechanism *is* the truth; I will merely argue that the epigenetic mechanism *may be* the truth. But the latter is important because it offers a possible (potentially testable) rationale for aspects of evolution, such as intelligence, where natural selection is wanting. Accepting the notion that the natural selection construct should be...and can be...expanded and improved constitutes the first step in furthering the field.

I will assume that there exist factors, over and above classical mutation-based natural selection, that have substantially upgraded our intelligence. One such "extra" factor might be the chemical communication between the brain and the germ cells (or their precursors). Chemical messengers could be produced at or near the site of the germ cells upon receipt of neurological signals from the brain or, more likely, the chemical messengers might be produced by the brain tissue itself. Either way, we know enough about gene plasticity and epigenetics to comprehend, in general terms, how the presence of externally produced substances can affect gene expression. This leads to the model's first postulate:

Postulate 1: Brain and germ cell tissues interact by means of chemical messengers.

There is, admittedly, no evidence that messengers produced in the brain, or in tissues stimulated by impulses from the brain, interact directly with the germ cells. All I offer (for the moment at least) is speculation, albeit speculation forced upon us by the inadequacies of natural selection as a comprehensive mechanism for human evolution. Speculation should not be confused, however, with meaningless proposals that are inherently unfalsifiable. I am not presenting here metaphysics beyond the possibility of scientific verification. For one thing, the construct predicts that one day chemical messengers will be identified. Although substantiating the existence of messengers in humans will no doubt prove difficult, this is an experimental problem, not a philosophical one.

It is not claimed, nor is it necessary for the ensuing model, that messengers originating from the brain interact only with the germ cells. Although a particular messenger might pervade all tissues, only the portion finding its way to the germ cells will have an effect on future generations. One cannot, of course, exclude the possibility that a given messenger does indeed concentrate in germ cells relative to other tissues. In this connection, I might mention a

Drosophila variant that produces elevated levels of aldehyde oxidase (an enzyme) in the male sex glands but nowhere else. By some unknown mechanism, perhaps via the intervention of an inducer or other control element, the switching on of the aldehyde-oxidase gene is tissue-specific.

The chemical identity of possible brain/germ-cell messengers is unknown and will remain so until someone manages to detect and isolate them. The messengers might be small molecules (no bigger than steroid hormones and neurotransmitters) or large molecules such as proteins or RNA. Brief mention should be made of a few of the widespread examples in biology where cells are profoundly affected by compounds produced at a source distant from those cells. I have already mentioned RNA interference in which small pieces of RNA can travel through multiple tissues to cause gene silencing (e.g. from the rootstock of a tobacco plant to a graft 30 cm away). Here are some other examples of cells being profoundly affected (genetically or otherwise) by substances produced at a source distant from those cells:

a) Insulin, secreted by the pancreas, enters liver cells and stimulates them to convert glycogen (animal starch) into glucose. This effect, induced by a compound synthesized far away from its site of action, happens not to be heritable as in the next example.
b) Cytoplasmic proteins and steroids can enter a nucleus, bind to DNA, and induce or suppress specific genes (see **Section 3**). Thus, compounds originally produced externally to a cell often have little trouble reaching the cell's DNA. This is further borne out by experiments showing that compounds of 5000 or less molecular weight, when injected into a cell's cytoplasm, appear virtually instantly within the nucleus. Even large proteins are known to diffuse through pores spanning the nuclear membrane at rates inversely proportional to their size.
c) Exposure of *E. coli* bacteria to salicin will accelerate the formation of a specific mutation or cause cells with that mutation to proliferate.

d) We have discussed with regard to the immune system how a bacterial antigen can induce the proliferation of only one lymphocyte cell among millions of closely related cells.
e) Many chemicals (hydrazines, mustards, aromatic amines, polycylic hydrocarbons, etc.) can penetrate cell membranes and ultimately induce somatic and germinal mutations. Even drugs can be dangerous in this regard. For example, thalidomide (a mild sedative no longer legally used in the United States) entered and damaged germ cells of patients who later had children with severe birth defects.

In summary, it is almost a trivial notion that both natural and synthetic compounds from external sources can enter cells and have a major effect upon their physiology and genetics. ***Postulate 1*** extends this behavior further: It states that brain cells can directly or indirectly initiate the manufacture of one or more compounds (messengers) that enter into and perturb the genetics of other tissues, particularly those of the reproductive organs.

Two possible mechanism are visualized by which the brain can influence the biochemistry of the germ cells: (a) stimulation of messenger production in or near the germ cells via nerve-transmitted signals from the brain and (b) manufacture of messenger by the brain itself followed by entry of the messenger into the blood stream and delivery to the germ cells. Let us consider the plausibility of each of these mechanisms.

No adult needs reminding that there exists a close connection between the brain and the sexual organs. It turns out that penile erection following tactile stimulation is mediated through the spinal cord. In contrast, erection following psychic stimulation depends upon cerebral centers in the brain. In the latter case, erotic thoughts or images cause pelvic neurons to initiate the production of nitric oxide (NO). Nitric oxide, a wonderfully simple compound, has been established as the principal physiological mediator of penile erection. It functions by increasing the formation of another chemical, abbreviated cGMP, which in turn allows blood to fill empty but expandable cavities, and an erection ensues.

Little wonder that the editors of *Science* in 1992 designated nitric oxide as the "Molecule of the Year". The point of this discussion is to demonstrate that brain activity can indeed stimulate the production of a bioregulatory molecule in the genital area. It is not too much of a stretch to presume that chemical messengers affecting DNA expression are induced at or near the germ cells by cranial signals.

But there is another possibility: The brain itself might produce the chemical messenger. "Preposterous," one might think, "the brain is a marvelous organ, but it is not a gland!" As it happens, the brain expresses a greater range of proteins than any other body tissue. Large numbers of proteins, some found only in specific locations or produced in such minute quantities they are barely detectable, are constantly being formed, broken down, and re-synthesized. The brain can also exude hormone-like chemicals. The process is called "neurosecretion", the products are called "neurotrophins", and the field of study dealing with this topic is called "neuroendocrinology".

Neurosecretory neurons can be defined as nerve cells that engage in secretory activity at a level comparable to gland cells. In primitive invertebrates, which lack well-developed glands or internal secretion, the nervous system is a main agency for carrying out the required endocrine functions. In the ganglia of round worms, for example, over one-half of all neurons are the neurosecretory type. Thus, neurohormones are not an evolutionary latecomer; they date back to the very beginnings of animals' neural development.

Experiments by I. R. Hagedorn on a species of leech (*Hiro medicinalis*) have shown that a hormonal influence from the brain controls the production of sperm. Thus, brain removal resulted in a severe depression both in the total number of germ cells and in the ability of germ cells to mature. Injection of brain tissue into these animals partially restored the spermatogenesis (in contrast to injection of muscle tissue that had no effect), strongly suggesting a brain-generated "gonadotropic" hormone.

The hypothalamus of humans, a small part of the brain weighing only 4 grams, is endowed with neurosecretory cells that manufacture various hormones such as somatostatin. Somatostatin is a 14-amino acid peptide that inhibits the release of a growth hormone from the nearby pituitary gland. Prior to release, neurohormones and neuronal chemical messengers are stored in hollow spheres called "neurosecretory vesicles". By means of the hormonal output of the hypothalamus, and its intimate relationship with the pituitary, the hypothalamus plays a major role in such activities as controlling emotional change, signaling hunger and thirst, and maintaining a constant internal environment ("homeostasis"). The hypothalamus is, in turn, responsive to numerous properties of the circulating blood including temperature, osmotic pressure, and the concentration of external hormones. For example, thermosensitive neurons activate vasodilatation and sweating when the blood temperature is elevated.

The term "*peptidergic neuron*" was assigned to neurons having microscopically visible granules of bioactive peptides that can serve as chemical messengers. At first it was thought that peptidergic neurons were a special type of neuron limited to a few tissues such as the hypothalamus. This is illustrated by a 1978 quote from B. Scharrer (a pioneer in the neurosecretion field): "Peptidergic neurons represent a highly specialized minority, capable of long distance (neurohormonal) as well as close range (neurohumoral) and intermediate forms of information transfer to various effector cells." Gradually, however, it was noticed that almost every neuron terminal contains at least a few large granules, hence the 1985 quote from T. Fujita: "Thus, it is now clear that essentially every neuron is more or less peptidergic" [peptide-producing]. "Moreover, all neurons cosecrete substances which belong to several chemical categories."

To summarize the last few paragraphs: Neurons of the brain are much more than mere transmitters of nerve impulses. They are chemical factories producing chemical messengers involved in complicated "feedback" mechanisms. This fact has a bearing on my theory for the evolution of human intelligence about to unfold.

I have mentioned that products of the brain can affect distant tissues including, in the case of leeches, the sperm cells. The reverse is also true, as shown by the action of liver-produced testosterone in songbirds such as the male canary and zebra finch. During breeding season when the males court the females, the blood level of testosterone reaches a peak. One effect of this testosterone is to increase the number, size, and complexity of neurons in the regions of the brain that are concerned with vocalization and aural interpretation of song. Injection of testosterone into the male birds out of the season, or into females, which have less developed song regions in their brain, causes a proliferation of "song" neurons and thus an anomalous appearance of song. Chemical brain/tissue communication obviously works in both directions.

Having affirmed the local and distant production of chemical messengers as a result of brain function, and having set forth the postulate that the germ cells are a possible receptor site for one or more messengers, it is now time to make a second assumption:

Postulate 2: Mental activity stimulates the production of messenger.

Biology is replete with examples of tissue activity stimulating specific chemical responses. Thus, muscle activity drives the formation of lactic acid; stressful activity raises the level of adrenaline; climbing to high altitudes causes red blood cells to synthesize D-2,3-bisphospoglycerate; nerve activity stimulates the production of acetylcholine; parturition (birth) triggers hormonal changes which in turn promote lactose synthesis needed for milk formation. It is obvious that brain activity (thinking, memorizing, problem solving, music playing, etc.) is also involved in a multitude of chemical changes accompanying the neural rewiring. The chemical changes can occur either within the brain or in non-cranial tissues to which the brain communicates via nerve impulses. I now propose that production of messengers, directed to the germ cells, constitutes one of the chemical changes that results from mental activity, a bold assertion without direct experimental evidence to date. Yet the reasonableness of the proposal has anecdotal support as seen below.

Research into the relationship between learning and chemistry has been hampered by the unfathomable complexity of the nervous system. Fortunately, for the purpose of this book I need not delve heavily into the neurochemistry of learning and memory. Brief examples will suffice because my only objective is to substantiate the fact that learning can indeed promote a cascade of interesting neurochemical events and that, therefore, the idea of learning-induced messengers is not to be dismissed out of hand.

Chemistry-of-learning experiments consist of three main types: (a) training an animal and then examining its brain for altered synthesis of various substances; (b) studying the effects of drugs on the ability of an animal to remember a given task; (c) administering brain tissue from a trained animal to an untrained animal in order to see if learning has been transferred. For ethical reasons, none of these approaches is suitable, of course, for human subjects which accounts, in part, for our current ignorance of human neurochemistry. In any event, let us consider typical examples of each type of experiment.

Rats were trained for four days to reverse their handedness, i.e. to obtain food with the paw opposite to their normal preference. The shift in handedness persisted about nine months following training. Brain cells from the side of the cortex believed to control the acquired handedness were then analyzed for RNA. As a control, cells from the opposite side of the cortex were similarly analyzed. It was found that only the "active" side of the cortex had an unusually high RNA content. Despite uncertainties in the data (arising from experimental error in the microanalyses, the lack of additional controls such as examination of tissues other than the brain, etc.), the conclusion from this and many related studies coincide: Certain types of training lead to changes in the content and structure of RNA at specific neurons of the brain. Further details can be found in H. Hydén and E. Egyhazi, *Proc Natl Acad Sci USA*, **53**, 946, 1965.

As an example of the drug approach to learning and memory, mention might be made of studies in mice by L. B. Flexner *et al.*

in *Science,* **155**, 1377, 1967. After mice were trained to run a maze, they were injected in the brain with puromycin (an antibiotic that inhibits protein synthesis). If puromycin is administered one to three days after training, the mice will not remember how to negotiate the maze. If the drug is given more than three days later, it has no effect on the training experience. One can conclude from studies such as this that protein synthesis plays a role in the consolidation of long-term memory, the exact mechanism of which is at present unclear.

Transfer-of-learning experiments, of which those of G. Ungar *et al.* in *Nature,* **238**, 198, 1972 were among the earliest, have received both wide publicity and controversy. Rats were trained to fear and avoid the dark, a response contrary to their normal behavior. When material from the brains of the trained rats was injected into untrained rats, the latter also exhibited fear of the dark. Brain material from untrained rats had no such effect. The active substance, isolated and purified from the brains of trained animals, was found to be a 15-amino acid peptide (a small protein) called "scotophobin" (derived from Greek words signifying darkness and fear). Scotophobin synthesized in the laboratory was reported to elicit the same fear-of-dark response as did biologically derived material. Despite appreciable skepticism surrounding the Ungar data, results corroborating the presence of learning-induced brain substances have been published by many laboratories.

In summary, there is no question that profound chemical changes occur in the brain during a learning experience. It is less clear exactly what component of the learning process gives rise to the chemical changes, and what ultimate purpose the newly formed molecules serve. Be that as it may, the data support the plausibility of ***Postulate 2***, namely that mental activity can stimulate chemical messengers in the brain and, possibly, in tissues outside the brain.

To recapitulate: ***Postulate 1*** says that brain and germ cell tissues interact via chemical messengers. And ***Postulate 2*** says that mental activity stimulates the production of such messengers. My third postulate, embodying the first two, finally reaches the heart of the matter:

Postulate 3: Chemical messengers, produced by mental activity, induce changes within the germ cells that are passed on to the next generation.

Assume, for example, that an external stimulus, such as a messenger from the brain, "turns on" a silent gene (or silent group of genes). No actual mutation occurs — just a change in gene expression. How can this imposed effect perpetuate itself? How can the imposed effect persist in later generations even in the absence of the stimulus? To address these questions, one need only postulate the following mechanism (among many other possible alternatives): The affected gene, once turned on, produces (directly or indirectly) a regulatory product that binds to the control region of the very same gene. The binding in turn activates the gene even in the absence of the original stimulus. In other words, since the gene is self-activating, once the gene becomes turned on by a cranial-based messenger the gene no longer requires this stimulus to escape its silence. This is epigenetics in action (consult *Chapter 17* for further details).

A fertilized ovum, acquiring some of the regulatory product from the cytoplasm of one or both germ cells, would possess the self-activated form of the gene even in the absence of messenger, and this trait would be passed on to daughter cells. One might recognize this mechanism from *Chapter 17* as a self-sustaining loop. The main point here is that a newly expressed gene appears on the scene as a result of a transient environmental stimulus (i.e. mental activity) and not as the result of a random mutation of DNA. The mechanism itself is not out-of-the-ordinary, being classically epigenetic in character, but the origin and focus of the environmental stimulus...the brain and the germ cells, respectively...is a novel and unproven idea.

For the sake of simplicity, a single gene was used to illustrate chemical communication between the brain and germ cells, but this grossly understates the reality that likely pervades the system. Since genes for complex traits are members of multiple gene networks, a perturbation by a single messenger derived from the brain can have a snowballing effect on the overall gene activity and the resulting phenotype. If, for example, a gene network is composed of one hundred genes (not an unreasonable number), then a messenger arousing a single gene from its silence could, possibly, cause several other genes to change their expression status in both directions. Such changes could have an appreciable effect on mental abilities, and this brings up the fourth and final postulate.

Postulate 4: The epigenetic changes in germ cells induced by mental activity facilitate mental activity in ensuing generations.

Postulate 4 closes the second of two self-sustaining loops as seen now in the following summary of the overall proposed mechanism: Mental activity produces chemical messengers that activate silent germ-cell genes, causing the genes to produce activators that turn on those genes that produced the activators in the first place (closing loop number 1). The genetic modifications are transmitted to embryonic cells and, as a consequence, neural networks formed during embryo development are modified to enhance the learning capabilities of the brain. The brain can now produce more messenger to continue and expand the process in future generations (closing loop number 2).

Let us examine the evolutionary consequences of this remarkable speculation:

Consider prehistoric humans living on the edge of survival as hunter-gatherers 50 000 years ago or 500 000 years ago (it does not matter which). A mutation appears that, by any number of possible mechanisms, enhances an individual's intelligence. Perhaps the number of neurons increases; perhaps the level of a neurotransmitter becomes elevated; perhaps there is an improved neuronal rewiring at an embryological stage. Whatever the exact mechanism for the improved intelligence, it likely entails an energy cost (organs-of-perfection, especially the brain, are very expensive items). Now there are two possibilities: (a) The individual with the enhanced intelligence might possess a reproductive advantage over his/her neighbors, and the trait will be spread throughout the population. This is conventional neo-Darwinism (with all its attendant problems that have already been amply discussed); (b) alternatively, the enhanced intelligence imparts no reproductive advantage at the time of its appearance. Or, if there is a functional advantage, it is more than nullified by the associated energy costs. With both these possibilities, either the trait will get "weeded out" or else the gene will be turned off (i.e. "masked" or "silenced") so that the gene no longer expresses itself. Energy costs of the useless gene are thereby eliminated. It is the masking process that is of interest.

Before continuing, I should pause and address again an obvious question: "How can an improvement in intelligence not be beneficial for survival?" In the area of linguistic intelligence, it is difficult to see how our brains capable of forming centers devoted to writing grammar can help a prehistoric person who does not write. In the area of musical intelligence, it is difficult to see how an ear of the type necessary to play the violin could be applied to hunting-gathering. And in the area of mathematical intelligence, it is difficult to see the evolutionary benefit of a brain that can understand something as abstract as a cubic root let alone a differential equation. One has to agree with Alfred Russell Wallace, the co-discoverer of evolution by natural selection, who believed that natural selection can account for most practical life-sustaining traits but certainly not for human cognitive abilities such as those mentioned above.

In the present theory, genes are deleted or masked because they possess no immediate advantage relative to their energy cost. Whether such a gene is deleted or masked depends in part upon the availability of a suitable masking chemistry. Undoubtedly, both deletion and masking occurred, but only the masked genes, lying cryptically in the vast sections of human DNA that do not manufacture proteins, have any future relevance. Thus, over the millennia we slowly accumulated genes that were never expressed but that, instead, represented a wealth of intelligence potential. Once released, our intellect appears as if we possess a trait totally out of synchrony with our activities during our evolutionary development. Recall that, in contrast, neo-Darwinism does not allow genes, and the traits they engender, to be out of step with the immediate demands of the environment. Mutations are retained only because they impart survival benefit to the organism at the time of their formation.

In order to complete the mechanism (which thus far has many of the components of what has been called "*neutral evolution*"), the masked intelligence genes must be switched on, and this is where ***Postulate 4*** comes into play. As humans challenged their brains with various thinking tasks, chemical changes were induced in the germ cells that, in a marvelous feedback process, unlocked the genetic potential that had amassed over the course of evolution. In prehistoric times, activities such as tracking game, making pottery, fashioning spears, and building fires helped unmask our potential. More recently, the fruits of formal education, among many other stimuli found in modern living, continue doing the same thing. Owing to the feedback aspect of the mechanism, intelligence has increased slowly over time but seemingly at a continuously increasing rate. This time-dependent unlocking expansion of our intellect explains (finally!) why we appear to be so much smarter now than the survival needs of prehistoric times dictated. We are, in fact, smarter now than in prehistoric times! But this has more to do with cultural reawakening of our genes than with neo-Darwinism based on rare and usually deleterious mutations.

The question is often debated as to whether human intelligence is cultural or genetic. We see now that the question makes no sense because human intelligence cannot be easily separated into two distinct cultural and genetic contributions. Cultural factors, including but certainly not limited to education, awaken a genome poised to expand our intellect. In other words, culture and genetics are intimately entwined at the DNA level. The implications of this model are clear: Mental activity in an adult prior to having children facilitates mental activity in the children of that adult. Stated another way, a child conceived after its parents have confronted mental challenges, including education and other life experiences, has overall a slightly higher probability (although no certainty!) of inheriting an improved learning ability.

Most serious scholars now believe that the intellectual faculties of humans have evolved as an adaptation to the complexities of social life (culture). Although this seems to be a valid enough portrayal of reality, the details of the mechanism are disconcertedly absent. Exactly how does a social life translate into an inherited adaptation? I have presented here a possible explanation for the interconnection between inheritance and culture as applied to intelligence and, until the time comes that the theory can be falsified, it must be considered a viable possibility.

It is necessary to immediately forestall any misunderstanding. Although mental activity is postulated as favoring the development of "smartness" genes in the progeny, this by no means signifies that high intelligence cannot arise from parents who, for example, have spent their lives at what some would consider "menial" work. (Quotes are placed around the word "menial" because, although the word is used widely, I for one dislike it; one should never underestimate the brainpower needed for competent pursuit of most vocations and avocations). Even the most "menial" occupation involves mental activities that far exceed the capabilities of the most intelligent non-human primates. You will never see a chimpanzee flipping hamburgers (a task commonly invoked as the epitome of humble work). Moreover, human intelligence is a complicated business and, obviously, it has many contributing factors in addition to epigenetics. Perhaps gene unmasking, if it does indeed contribute to the overall picture, can also occur spontaneously without any prodding via actual mental activity. Certain gene combinations, arising from sexual reproduction, may allow relevant genes to turn on and off apart from epigenetic factors. On the other side of the coin, perhaps a particular child will fail to benefit from previous mental activity of its parents owing to other overriding influences including cultural factors (e.g. a poor home-life) and genetic factors (e.g. inheritance of unfavorable genes). The point is that we are dealing here with population genetics. If, however, epigenetics plays a role in the evolution of intelligence, then children will, on the average, and when taken over the entire population over extended periods of time, become smarter as their parents are more and more engaged in mentally challenging activities prior to their reproductive years.

Mental challenges need not necessarily be of the academic type. To borrow from a previous statement in this book: Reading a newspaper, fixing a toy, driving a car, operating a word processor, caring for a pet, admiring a flower, solving a crossword process, balancing a checkbook, playing a game of chess, filling out a tax form, planning a vacation, painting a picture, cooking a dinner, confessing a sin, reflecting on God, and predicting future events, all engage our intellect. But among such activities, one would guess that formal schooling is a particularly efficient route to heritable mental development.

The foregoing "*cranial feedback mechanism*" has, admittedly, a distinct Lamarckian touch to it (just as there is a Lamarckian touch to epigenetics and biotechnology). But the mechanism must never be extrapolated to Lamarckian extremes such as, for example, a need-driven production of webbed feet in birds that live on water. For one thing, there is no reason to postulate feedback between bird feet and the genome. Webbed duck feet are perfectly in tune with environmental requirements. So Darwinism works reasonably well here (the obscurity of the details notwithstanding). Such is not the case with human's mental capacity which, I have argued at great length, demands an alternative rationale of which the cranial feedback mechanism is one possibility.

The cranial feedback mechanism has substantial differences and advantages over classical neo-Darwinism:

1. Epigenetic modification is much faster than modification by mutation. Only epigenetics can, therefore, explain the rapid development of human intellectual accomplishments over the past 50 000 years. Only epigenetics can explain why intelligence is uniformly distributed worldwide because no mixing of "intelligence genes" over vast geographical areas is required.
2. Epigenetic modification responds to actual needs in contrast to mutations that are random and most frequently deleterious.
3. Only epigenetic modification is consistent with the obvious difference in the mental capabilities between the chimpanzee and human genome despite their 99% similarity in genomes.

In connection with the third point, it should be mentioned that comparison of cerebral cortex samples from humans and chimpanzees have identified 83 genes in humans that are more active than the corresponding genes in chimpanzees. Gene activities in other tissues (e.g. heart and liver) do not show this effect. Thus, humans and chimpanzees differ not so much in the identities of the genes controlling neural activity but in the level of gene activity. This, of course, lends itself to epigenetic control.

There is no reason to believe that we have reached our intellectual zenith. If a degree of genetic potential for additional mental prowess lies hidden within the human genome, then one can make a reasonable prediction: Humans will become smarter and smarter. And this process will be aided, one would expect, by educational systems that truly challenge children to employ their mental facilities, in which case the children of these children will have (on the average taken over the entire population with, as noted, frequent exceptions) an easier time of it. Teachers may take comfort in the possibility that, by forcing students to study and learn, more than one generation is being affected. Universal education has just begun, relative to the evolutionary time-scale, and its effects might likely become more and more manifest as the centuries progress. If this is correct, then establishing high-quality educational systems in societies throughout the world should be a high-priority goal of humanity.

Psychologist James Flynn carried out extensive surveys showing that mean IQ scores of groups all around the world are steadily increasing, decade after decade (e.g. a 15-point increase over two or three decades is not unusual). In fact, the increase has become known as the "Flynn effect". As stated in the early part of this book, IQ test scores must be treated with great caution, and I have generally avoided them. Thus, the Flynn effect could be caused, for example, by improved test-taking skills, more nutritious diets, better textbooks, or easier exams (or, in other words, "culture"). Owing to these ambiguities, the Flynn effect, although consistent with the cranial feedback mechanism, cannot be strongly invoked to support it.

One of the most common questions voiced by students and the public is, "Are humans still evolving?" We have argued above that, contrary to the traditional viewpoint, humans of today are not a finished product. In fact, Dan Dediu and Robert Ladd of the University of Edinburgh believe in a trend toward a quickly accelerating evolution in humans. Gregory Cochran of the University of Utah talk of a "fantastically rapid" recent human evolution. Now consider the fact that in my state of Georgia, the birth rate among poor women, and those with less than 12 years of education, is three times that of the rest of the population. It seems unlikely that these reproductively successful women, with a Darwinian advantage over everyone else, have an average intelligence higher than the average intelligence of the general population. Thus, it is difficult to see how a classical Darwinian natural selection could be at play here. Where, for one thing, does natural selection enter the picture? Some other factor or factors must be making us more "brainy", and epigentics seems to be a likely candidate.

A modern chemist, as I can affirm from personal experience, is a storehouse of facts, concepts, and theories devised by others...past and present; such information has accumulated over the years until chemists now have a command of the field that would amaze those of only a decade or two ago. There is no denying, therefore, that culture provides a necessary component of chemists' "intelligence". But "*The Thin Bone Vault*" does not address this aspect of "cultural intelligence" except, perhaps, to laud the fact that the human brain is capable of absorbing all that chemical information. More relevant are other aspects of a chemist's mind: A capacity for thinking abstractly, for reasoning, for predicting, and for organizing large quantities of abstruse information into meaningful systems; an ability to be imaginative with symbols and equations; a skill and intuition in generating creative solutions to difficult problems; a sense for developing new combinations of old ideas and for guessing complicated interrelationships; a talent for devising new experiments that shed light on chemical mysteries; an alertness for the out-of-the-ordinary that might lead to the development of a new theory or principle. A brain that can do all this is not a brain that was needed by prehistoric man. It is a brain that developed rather quickly within the past 50 millennia (which is far too short a time period to invoke an origin by point-mutational changes). Cultural advances must have unlocked genetic potential in a cranial feedback mechanism which, one might reasonably propose, continues to this present day. Culture provides us with raw information and know-how, but culture may also endow us with the ability to pass on to our heirs an intellect capable of absorbing and manipulating this information and know-how.

The model assumes an amazingly malleable ("*plastic*") human genome — not a genome burdened by an excrutiatingly slow (and usually deleterious) dependence upon point-by-point mutational change in order to achieve a measure of improvement. It should be clear now why I have previously discussed other examples of gene plasticity: (a) snapdragons in which low temperature favors a transposon relocation and consequent ivory-to-red color change; (b) sleeping sickness in which a parasite changes its gene expression, and resulting antigenic proteins, every week or two;

(c) the immune system where each lymphocyte selects its own combination of genes with which to create a structurally unique antibody; (d) "heat shock" genes in *E. coli* that produce 17 new proteins upon exposure to high temperature. Although these processes may or may not have a direct bearing on human genetics, they do demonstrate a far greater potential for rapid genetic environmentally-driven changes than is involved in classical neo-Darwinism.

There are, of course, uncertainties with the cranial feedback mechanism. How is it, for example, that masked genes remained unaltered for long periods of time under the mutational pressure endured by all genes? It may be, in answer to this question, that the masked intelligence genes are not very old and, therefore, have not had much time to mutate; that repair mechanisms have helped stabilize the masked genes; that gene duplication has to some extent protected the genes from mutational degradation. A better response is that we simply do not know the answer. We also do not know how many generations the epigenetic improvement in intelligence would persist should intellectual stimuli suddenly cease. But ignorance at this stage is no cause for shame or alarm. When Darwin wrote his great works, the field of genetics was yet to be developed, and, as a consequence, for decades we had to endure major gaps in our understanding of his theory. I am in a similar position owing to the fact that the field of epigenetics is in its infancy. I can only content myself with the hope that, despite many uncertainties, the foregoing epigenetic theory of intelligence (or something akin to it) will stimulate attempts to disprove its validity. This is an important challenge because as Steven Pinker stated: "The apparent evolutionary uselessness of human intelligence is a central problem of psychology, biology, and the scientific worldview."

Apart from proving or disproving any theory, however, I have had a broader goal, and in this regard I echo the sentiments of Mae-Wan Ho in his book "*The Rainbow and the Worm*":

> To me, science is a quest for the most intimate understanding of nature. It is not an industry set up for the purpose of validating

existing theories and indoctrinating students in the correct ideologies. It is an adventure of the free, enquiring spirit which thrives not so much on answers as unanswered questions. It is the enigmas, the mysteries, and paradoxes that take hold of the imagination, leading it on the most exquisite dance. I should be more than satisfied if, at the end of this book, I have done no more than keep the big question alive.

Section 5

BIBLIOGRAPHY

The following books listed proved useful in the preparation of this book. Listing of a book does not imply the author's agreement with its theme. Since the literature on evolution is vast, the compilation of books is necessarily incomplete. Within the books are citations to other references that suitably blanket the field.

Anfinsen, C. B., *The Molecular Basis of Evolution*, Wiley, 1959.

Appleman, P. (Ed.), *Darwin*, W. W. Norton, 1979.

Atkins, P. W., *The Second Law,* Scientific American Books, 1984.

Aunger, R., *The Electric Meme: A New Theory of How We Think*, Free Press, 2002.

Behe, M. J., *Darwin's Black Box*, Free Press, 1996.

Behe, M. J., *The Edge of Evolution. The Search for the Limits of Darwinism*, Free Press, 2007.

Berlinski, D., *Black Mischief*, Harcourt Brace Jovanovich, 1988.

de Boysson-Bardies, B., *How Language Comes to Children*, MIT Press, 1999.

Brockman, J. (Ed.), *Creativity*, Simon & Schuster, 1993.

Campbell, J., *The Improbable Machine*, Simon & Schuster, 1989.

Casti, J. L., *Paradigms Lost*, William Morrow and Co., 1989.

Coppinger, R. and Coppinger, L., *Dogs: A Startling New Understanding of Canine Origin, Behaviour and Evolution*, Scribner, 2001.

Corballis, M. C., *From Hand to Mouth: The Origins of Language*, Princeton University Press, 2002.

Davies, P., *The Cosmic Blueprint*, Simon & Schuster, 1989.

Dawkins, R., *The Blind Watchmaker*, W. W. Norton & Co., 1987.

Dembski, W. A., *No Free Lunch: Why Specified Complexity Cannot Be Purchased Without Intelligence*, Rowman & Littlefield, 2002.

Dennis, C. and Gallagher, R. (Eds.), *The Human Genome*, Palgrave, 2001.

Denton, M., *Evolution: A Theory in Crisis*, Adler & Adler, 1986.

Devlin, K., *Goodbye Descartes*, Wiley, 1997.

Donald, M., *Origins of the Modern Mind: Three Stages in the Evolution of Culture and Cognition*, Harvard University Press, 1991.

Donald, M., *A Mind So Rare: The Evolution of Human Consciousness*, W. W. Norton, 2001.

Dugatkin, L. A., *The Imitation Factor: Evolution Beyond the Gene*, Free Press, 2001.

de Duve, C., *Blueprint for a Cell: The Nature and Origin of Life*, Neil Patterson, 1991.

Ebon, M. (Ed.), *Psychic Discoveries by the Russians*, New American Library, 1963.

Eccles, Sir J. and Robinson, D. N., *The Wonder of Being Human*, Shambhala, 1985.

Edelman, G. M., *Bright Air, Brilliant Fire*, Basic Books, 1992.

Ehrlich, P. R., *Human Natures*, Island Press, 2000.

Eldredge, N., *The Monkey Business*, Washington Square Press, 1982.

Eldredge, N., *The Triumph of Evolution and the Failure of Creationism*, Macmillan, 2000.

Fragaszy, D. M. and Perry, S. (Eds.), *The Biology of Traditions*, Cambridge University Press, 2003.

Gardner, H., *Frames of Mind*, Basic Books, 1983.

Gerard, R. W., *Unresting Cells*, Harper & Brothers, 1940.

Godfrey, L. R. (Ed.), *Scientists Confront Creationism*, W. W. Norton & Co., 1983.

Goldschmidt, T., *Darwin's Dreampond*, MIT Press, 1996.

Gould, S. J., *Ever Since Darwin*, W. W. Norton & Co., 1977.

Gould, S. J., *An Urchin in the Storm*, W. W. Norton & Co., 1987.

Gould, S. J., *Bully for Brontosaurus*, W. W. Norton & Co., 1991.

Hawkins, J. and Blakeslee, S., *On Intelligence*, Times Books, 2004.

Hayward, A., *Creation and Evolution*, Bethany House, 1995.

Ho, M-W., *The Rainbow and the Worm*, World Scientific, 1993.

Hofstadter, D. R., *Godel, Escher, Bach: an Eternal Golden Braid*, Vintage Books, 1980.

Horgan, J., *The End of Science*, Addison-Wesley, 1996.

Huxley, J., *The Living Thoughts of Darwin*, Fawcett, 1959.

Jablonka, E. and Lamb, M. J., *Evolution in Four Dimensions*, MIT Press, 2005.

Jackendoff, R., *Foundations of Language: Brain, Meaning, Grammar, and Evolution*, Oxford University Press, 2002.

Johnson, P. E., *Darwin on Trial*, Regnery Gateway, 1991.

Jones, S., Martin, R. and Pilbeam, D. (Eds.), *The Cambridge Encyclopedia of Human Evolution*, Cambridge University Press, 1992.

Kappel-Smith, D., *Night-Life*, Little, Brown, & Co., 1990.

Kauffman, S., *At Home in the Universe*, Oxford University Press, 1995.

Klein, J. and Takahata, N. *Where Do We Come From? The Molecular Evidence for Human Descent*, Springer, 2002.

Konner, M., *The Tangled Wing*, Henry Holt and Co., 1982.

Lakoff, G. and Núñez, R. E., *Where Mathematics Comes From: How the Embodied Mind Brings Mathematics Into Being*, Basic Books, 2000.

Leakey, R. E. and Lewin, R., *People of the Lake*, Avon Books, 1978.

LeDoux, J., *Synaptic Self: How Our Brains Become Who We Are*, Pan Macmillan, 2002.

Lewin, R., *Human Evolution*, Blackwell Scientific Publications, 1984.

Lewin, R., *Bones of Contention*, Simon & Schuster, 1987.

Lewontin, R. C., *Biology as Ideology; The Doctrine of DNA*, Harper Perennial, 1992.

Lieberman, P., *Human Language and Our Reptilian Brain: The Subcortical Bases of Speech, Syntax, and Thought*, Harvard University Press, 2000.

Linden, D. J., *The Accidental Mind: How Brain Evolution Has Given Us Love, Memory, Dreams, and God*, Belknap Press, 2007.

Luria, A. R., *The Mind of a Mnemonist*, Henry Regnery Co., 1968.

Margulis, L. and Sagan, D., *Acquiring Genomes: A Theory of the Origin of Species*, Basic Books, 2002.

Matsuzawa, T. (Ed.), *Primate Origins of Human Cognition and Behavior*, Springer, 2001.

Mayr, E., *What Evolution Is*, Basic Books, 2001.

McCrone, J., *The Ape That Spoke*, William Morrow and Co., 1991.

McEachern, R. H., *Human and Machine Intelligence: An Evolutionary View*, R & E Publishers, 1993.

McKibben, B., *The End of Nature*, Doubleday, 1989.

Medina, J., *Brain Rules*, Pear Press, 2008.

Mithen, S., *The Prehistory of the Mind: The Cognitive Origins of Art, Religion, and Science*, Thames and Hudson, 1996.

Ofek, H., *Second Nature: Economic Origins of Human Evolution*, Cambridge University Press, 2001.

Ornstein, R. and Thompson, R. F., *The Amazing Brain*, Houghton Mifflin, 1984.

Palumbi, S. R., *The Evolution Explosion: How Humans Cause Rapid Evolutionary Change*, W. W. Norton, 2001.

Parker, S. (Ed.), *The Dawn of Man*, Crescent Books, 1992.

Penrose, R., *Shadows of the Mind*, Oxford University Press, 1994.

Pfenninger, K. H. (Ed.), *The Origins of Creativity*, Oxford University Press, 2001.

Pigliucci, M., *Phenotypic Plasticity: Beyond Nature and Nurture*, Johns Hopkins University Press, 2001.

Reid, R. G. B., *Evolutionary Theory: The Unfinished Synthesis*, Cornell University Press, 1985.

Relethford, J., *The Human Species*, Mayfield Publishing Co., 1990.

Rhodes, F. H. T., *Evolution*, Golden Press, 1974.

Richards, R. J., *Darwin and the Emergence of Evolutionary Theories of Mind and Behavior*, University of Chicago Press, 1987.

Richardson, K., *The Making of Intelligence*, Columbia University Press, 2002.

Rifkin, J., *Algeny*, Penguin Books, 1984.

Rose, S., *The Future of the Brain*, Oxford, 2005.

Sagan, C., *The Dragons of Eden*, Ballantine Books, 1977.

Sagan, C., *The Demon-Haunted World*, Ballantine Books, 1996.

Sagan, C. and Druyan, A., *Shadows of Forgotten Ancestors*, Random House, 1992.

Serebriakoff, V., *The Future of Intelligence*, Parthenon Publishing Group, 1987.

Shafto, M. (Ed.), *How We Know*, Harper & Row, 1985.

Sheldrake, R., *The Presence of the Past*, Times Books, 1988.

Skoyles, J. R. and Sagan, D., *Up from Dragons*, McGraw Hill, 2002.

Smith, J. M., *Did Darwin Get It Right?*, Chapman and Hall, 1989.

Steele, E. J., *Somatic Selection and Adaptive Evolution*, University of Chicago Press, 1979.

Sternberg, R. J. and Kaufman, J. C. (Eds.), *The Evolution of Intelligence*, Lawrence Erlbaum Associates, 2002.

Strogatz, S. H., *Nonlinear Dynamics and Chaos*, Addison-Wesley, 1994.
Taylor, G. R., *The Great Evolution Mystery*, Harper & Row, 1983.
Tinbergen, N., *Animal Behavior*, Time Incorp., 1965.
Van Doren, C., *A History of Knowledge*, Ballantine Books, 1991.
Wells, J., *Icons of Evolution: Science or Myth? Why Much of What We Teach about Evolution is Wrong*, Regnery, 2000.
Wenger, W., *How to Increase Your Intelligence*, Dell, 1975.
Wesson, R., *Beyond Natural Selection*, MIT Press, 1991.
West, R. M. E. (Ed.), *The Best of American Science Writing*, Ecco Press, 2000.
Wilber, K., *Up From Eden*, Shambhala, 1986.
Wilford, J. N., *The Riddle of the Dinosaur*, Vintage Books, 1985.
Wills, C., The *Wisdom of the Genes*, Basic Books, 1989.
Wills, C., *The Runaway Brain*, Basic Books, 1993.
Wilson, E. O. (Ed.), *From So Simple a Beginning: The Four Great Books of Charles Darwin*, W. W. Norton & Co., 2006.
Wilson, F. R., *The Hand*, Pantheon Books, 1998.

[illegible] *New [illegible]*, [illegible] Addison-Wesley, 199[illegible]
[illegible], C. [illegible] *[illegible] Evolution [illegible]*, Harper & Row, 196[illegible]
[illegible] *[illegible]*, [illegible] 196[illegible]
[illegible] *[illegible]*, [illegible] Books, 1991
[illegible] *The [illegible] of What [illegible]*
[illegible] 200[illegible]
[illegible], W. [illegible], Dell, 1[illegible]
[illegible] *[illegible]*, [illegible] Press, 195[illegible]
[illegible] *The [illegible]*, [illegible] 19[illegible]
[illegible]
[illegible]

INDEX